Petra M. Dorfner

Coated Implant Surfaces

Petra M. Dorfner

Coated Implant Surfaces

Influence of RGD Coating in Combination with Surface Properties on the Adhesion Process of Osteoblasts

Südwestdeutscher Verlag für Hochschulschriften

Impressum / Imprint

Bibliografische Information der Deutschen Nationalbibliothek: Die Deutsche Nationalbibliothek verzeichnet diese Publikation in der Deutschen Nationalbibliografie; detaillierte bibliografische Daten sind im Internet über http://dnb.d-nb.de abrufbar.

Alle in diesem Buch genannten Marken und Produktnamen unterliegen warenzeichen-, marken- oder patentrechtlichem Schutz bzw. sind Warenzeichen oder eingetragene Warenzeichen der jeweiligen Inhaber. Die Wiedergabe von Marken, Produktnamen, Gebrauchsnamen, Handelsnamen, Warenbezeichnungen u.s.w. in diesem Werk berechtigt auch ohne besondere Kennzeichnung nicht zu der Annahme, dass solche Namen im Sinne der Warenzeichen- und Markenschutzgesetzgebung als frei zu betrachten wären und daher von jedermann benutzt werden dürften.

Bibliographic information published by the Deutsche Nationalbibliothek: The Deutsche Nationalbibliothek lists this publication in the Deutsche Nationalbibliografie; detailed bibliographic data are available in the Internet at http://dnb.d-nb.de.

Any brand names and product names mentioned in this book are subject to trademark, brand or patent protection and are trademarks or registered trademarks of their respective holders. The use of brand names, product names, common names, trade names, product descriptions etc. even without a particular marking in this works is in no way to be construed to mean that such names may be regarded as unrestricted in respect of trademark and brand protection legislation and could thus be used by anyone.

Coverbild / Cover image: www.ingimage.com

Verlag / Publisher:
Südwestdeutscher Verlag für Hochschulschriften
ist ein Imprint der / is a trademark of
AV Akademikerverlag GmbH & Co. KG
Heinrich-Böcking-Str. 6-8, 66121 Saarbrücken, Deutschland / Germany
Email: info@svh-verlag.de

Herstellung: siehe letzte Seite /
Printed at: see last page
ISBN: 978-3-8381-3510-6

Zugl. / Approved by: München, TU, Diss., 2012

Copyright © 2012 AV Akademikerverlag GmbH & Co. KG
Alle Rechte vorbehalten. / All rights reserved. Saarbrücken 2012

TECHNISCHE UNIVERSITÄT MÜNCHEN

Institut für Organische Chemie und Biochemie

Coated Implant Surfaces - Influence of RGD Coating in Combination with Surface Properties on the Adhesion Process of Osteoblasts

Petra M. Dorfner

Vollständiger Abdruck der von der Fakultät für Chemie der Technischen Universität München zur Erlangung des akademischen Grades eines Doktors der Naturwissenschaften (Dr. rer. nat.) genehmigten Dissertation.

Vorsitzender	Univ.-Prof. Dr. St. J. Glaser
Prüfer der Dissertation	1. Univ.-Prof. Dr. Dr. h.c. H. Kessler
	2. Priv.-Doz. Dr. R. H. H. Burgkart

Die Dissertation wurde am 02.07.2012 bei der Technischen Universität München eingereicht und durch die Fakultät für Chemie am 19.07.2012 angenommen.

For Maximilian

CONTENT

PREFACE & ACKNOWLEDGMENT

ABSTRACT

ABSTRACT GERMAN

PUBLICATION, POSTERS AND ORAL COMMUNICATIONS

LIST OF ABBREVIATIONS

MOTIVATION

1 INTRODUCTION & BACKGROUND 1
1.1 Optimization of bone implants by improvement of osseointegration 1
1.2 Osteoblast adhesion is a complex process 7
   1.2.1 Bone structure and role of osteoblasts 7
   1.2.2 Osteoblast adhesion and the influence of integrins 10
   1.2.3 Different approaches to measure cell adhesion 15
1.3 Biofunctionalization with RGD peptides is a promising method for optimizing the selective adhesion 17

2 RESULTS & DISCUSSION 23
2.1 Primary osteoblasts were obtained from human bone 24
   2.1.1 Extraction and culturing of human primary osteoblasts 25
   2.1.2 Qualitative characterization of osteoblasts 26
   2.1.3 Confirmation of $\alpha v \beta 3$ expression on osteoblast membrane 32
2.2 Investigated surfaces vary in terms of roughness 33
2.3 Binding assays with Integrins verify affinity of RGD peptides 38
   2.3.1 Optimization of integrin binding assay for $\alpha v \beta 3$ 38
   2.3.2 $\alpha v \beta 3$ binding on RGD coated trimmed Ti6Al4V disks 44
   2.3.3 $\alpha v \beta 3$ binding on RGD coated Ti6Al4V disks with different surface roughness 46
   2.3.4 Quantitative analysis of RGD peptides bound on Ti6Al4V disks by radio labeling 48

2.3.5	Development of the assay for α5β1	51
2.4	**Impact of RGD peptides and surface structure on the osteoblast adhesion process**	**54**
2.4.1	RGD peptide concentration for cell assays was optimized	55
2.4.2	RGD peptides coated on Ti6Al4V disks are non-toxic for osteoblasts	56
2.4.3	Cilengitide inhibits binding of osteoblasts to Ti6Al4V disks	57
2.4.4	Cells prefer rough surfaces coated with RGD peptides for adhesion	59
2.4.5	BSA blocking of Ti6Al4V disks influences osteoblast adhesion	61
2.4.6	Amount of FCS in media affects the adhesion process	64
2.4.7	SEM images show spreading of cells in detail	68
2.4.8	Tracking of spreading shows effect of peptide and FCS on glass surface	70
2.4.9	Adhesion process is highly time dependent	74
2.5	**Surface structure is more important than surface treatments - demonstrated in a sheep experiment**	**80**
3	**SUMMARY**	**87**
4	**MATERIALS & METHODS**	**90**
4.1	**Materials**	**90**
4.1.1	Chemicals	90
4.1.2	Equipment	92
4.1.3	Disks	94
4.1.4	Cell culture and Immunochemistry	94
4.2	**Methods**	**97**
4.2.1	Roughness measurements of Ti6Al4V disks	97
4.2.2	Peptide coating of Ti6Al4V disks	97
4.2.3	vβ3-adhesion assay	97
4.2.4	α5β1-adhesion assay	98
4.2.5	Obtaining and culturing primary human osteoblasts	98
4.2.6	Culturing of human fibroblasts (HFIB)	99
4.2.7	Cell handling	99
4.2.8	Immunohistology protocol	100
4.2.9	Lactate dehydrogenase (LDH)	101

4.2.10	Evaluation of osteoblast adhesion by chromogenic hexosaminidase test	102
4.2.11	Fluorescent staining	102
4.2.12	Video spreading	102
4.2.13	SEM images	103
4.2.14	Sheep experiment	103
4.2.15	Statistical analysis	105

LITERATURE **106**

PREFACE & ACKNOWLEDGMENT

The present work was accomplished in the time from October 2007 till June 2012 within the International Graduate School of Science and Engineering at Technische Universität München (TUM).

In the beginning I would like to thank my advisors, Prof. Horst Kessler and PD Rainer Burgkart, for giving me the possibility to join this IGSSE project and work within their groups. They supported me continuously over the last years. It was not only the granting of very good working conditions, but also the possibility to develop own ideas, the constant willingness for discussion and the critical review that helped me to learn so much and to achieve the following doctoral thesis.

A special and warmest thank goes to Dr. Carles Mas-Moruno who was the best project team leader this project could get. I am deeply thankful that he joined Prof. Kessler's group and I got the chance to work with him. It was a pleasure to do experiments with him, I learned so much during our discussions and we became good friends.

The work in the lab is not always straight forward and can even get frustrating; therefore it was a pleasure to have positive, motivated, friendly and cooperative colleagues around. I like to thank Flo, Steffi, Monica, and Florian from the Prof. Kessler lab, Jutta, Belma, Sarah, Johannes, Carmen, Jochen, Vroni, Angelika, Matthias, Natascha and the guys from Merzendorfer's in the University Hospital Klinikum rechts der Isar.

To answer some interdisciplinary questions, I got supported by people around medicine, chemistry and also around university: With Ute's help I could prove and determine the presence of integrins on the cell membrane, Pablo showed me how to illustrate the adhesion process over the time by capturing our osteoblasts, Maximilian helped me by programming a tool to analyze the cell spreading, and Klaus from the Deutsche Museum was a great support to obtain the SEM-images of our samples

Last, but not least the greatest support came from family and friends. I'd like to thank Dominic for being understanding and pushy at the right time. Maximilian, who joint only the last two years, was the best child one could imagine and at one time he has to read what his mum has written here.

Munich, 20. August 2012   Petra Dorfner

ABSTRACT

A profound knowledge of bony ingrowth represents the basis for the development of novel biofunctionalized implants that are highly attractive and selective for osteoblast adhesion. The positive influence of peptide coating and surface properties of implant material on this adhesion are well known. However, these two factors are rarely studied in combination and, in particular, not in depth during the early phase of the adhesion.

With the aim to reveal potential synergies, we classified Ti6Al4V disks, a common material for cement-free implants, into different roughness categories (trimmed, matt finished and sandblasted) and functionalized them with highly active αv-specific RGD peptides. Our integrin binding studies yielded the most promising results in terms of $\alpha v \beta 3$ binding for the combination of peptide coating and sandblasted Ti6Al4V surfaces. The relative impact of peptide coating was strongest on trimmed surfaces. This observation can be explained with the better accessibility of peptides for integrins on smooth surfaces as opposed to topologically rougher surfaces. Our cell adhesion experiments were performed with a characterized cell pool of human primary osteoblasts. We demonstrated that also osteoblasts adhere more strongly on rough surfaces coated with RGD peptides compared to smooth surfaces. However, the positive effect of RGD peptide coating on the adhesion process can be impaired by the presence of serum, or other proteins, like bovine serum albumin (BSA). Both aspects have a critical influence on osteoblast adhesion and need to be considered in the design of cell adhesion studies. In addition, we showed that the effect of peptide coating and surface topography influences the cell adhesion kinetics particularly at the early stage. The spreading of osteoblasts becomes significantly accelerated by the presence of RGD peptides immobilized on Ti6Al4V disks within the first hour. After longer incubation times the promoting influence of peptide coating becomes less pronounced. The benefit gained through escalated osteoblast adhesion within the first hours is particularly relevant for a stable and selective osteoblast-implant interphase. This is of utmost importance, because only osteoblasts facilitate a stable osseointegration.

In summary, the combination of RGD peptide coating with optimized topological surface properties of implant materials bears the potential to increase osseointegration and, hence, to be beneficial for numerous patients.

ABSTRACT GERMAN

Ein grundlegendes Wissen über das knöcherne Einwachsen von Implantaten ist die Basis für eine gezielte Weiterentwicklung von neuen biofunktionalisierten Materialien, die hoch attraktiv und selektiv für Osteoblasten sind. Der positive Einfluss von Peptidbeschichtungen und Oberflächeneigenschaften auf das zelluläre Adhäsionsverhalten ist bereits gut bekannt. Diese beiden Faktoren wurden jedoch selten zusammen untersucht und Studien besonders während der Anfangsphase der Adhäsion sind rar.

Mit dem Ziel mögliche Synergien zu entdecken, haben wir Ti6Al4V-Plättchen, ein häufig verwendetes Implantatmaterial, mit verschiedenen Rauigkeiten (abgedreht, glasperlenmattiert, sandgestrahlt) klassifiziert und mit hoch aktiven αv-selektiven RGD-Peptiden beschichtet. Die beste $\alpha v\beta 3$-Integrinbindung konnten wir für die Kombination von Peptidbeschichtung und sandgestrahltem Ti6Al4V nachweisen. Dabei ist zu erwähnen, dass die relative Steigerung der Integrinbindung durch die Peptidbeschichtung auf abgedrehten (glatten) Oberflächen am größten war. Diese Beobachtung erklären wir mit der besseren Verfügbarkeit von Peptiden auf glatten Oberflächen im Vergleich zu rauen. Die zellulären Experimente wurden mit einem charakterisierten Zellpool von humanen primären Osteoblasten durchgeführt. Hier haben wir durch die RGD-Peptidbeschichtung eine stärkere Adhärenz von Osteoblasten auf rauen gegenüber glatten Oberflächen nachgewiesen. Jedoch kann der positive Effekt der RGD-Peptide auf den Adhäsionsprozess durch die Präsenz von Serum oder Proteinen, wie bovines Serumalbumin, abgeschwächt werden. Beides hat einen erheblichen Einfluss auf die Adhäsion und muss für die Planung von Zellstudien berücksichtigt werden. Zusätzlich wurde der Einfluss von Peptidbeschichtung und Oberflächentopographie auf den zeitlichen Ablauf der Zelladhäsion untersucht, speziell in der frühen Phase. Die Ausbreitung von Osteoblasten wird gerade in der ersten Stunde durch die Präsenz von Peptiden beschleunigt. Nach längerer Inkubationszeit wird die Wirkung der Peptide von anderen Effekten überlagert. Allerdings ist der positive Effekt, der durch diese gesteigerte Osteoblastenadhäsion zu Beginn erreicht wird, für eine selektive Osteoblasten-Implantat-Verbindung relevant, denn nur Osteoblasten sorgen für eine stabile Osseointegration.

Zusammenfassend stellt sich heraus, dass die Kombination von RGD-Peptidbeschichtung und optimierten Oberflächeneigenschaften des Implantatmaterials ein großes Potential darstellen die Osseointegration zu verbessern. Dadurch könnte vielen Patienten geholfen werden.

PUBLICATION, POSTERS AND ORAL COMMUNICATIONS

Publication

C. Mas-Moruno, P. Dorfner, F. Manzenrieder, S. Neubauer, U. Reuning, R. Burgkart, H. Kessler. Factors controlling osteoblast adhesion to Ti6Al4V: the role of peptide coating and surface roughness, *J. Biomed Mat Res Part A.* **2012**, 00A:000–000.

Poster presentations

P. Kleiner*, J. Branch, J. Bushman, J. Kohn, R. Burgkart. qPCR as a screening method to investigate the influence of surface features and soluble factors in matters of Osteoblast differentiation for hMSCs, *16th Swiss Conference on Biomaterials.* May 5, **2010**. Dübendorf (Switzerland)

F. Rechenmacher, S. Neubauer, C. Mas-Moruno, P. M. Kleiner*, R. Burgkart, H. Kessler. Derivatization of Selective Integrin Ligands via Click-Chemistry for Surface Coating. *4th IGSSE (International Graduate School of Science and Engineering) Forum.* Jun 14-18, **2010**. Raitenhaslach, Burghausen (Germany)

S. Neubauer, F. Rechenmacher, B. Laufer, C. Mas-Moruno, A. Bochen, P. M. Kleiner*, A. Schwede, J. P. Spatz and H. Kessler. Highly Active and Selective Integrin Ligands and Their Application for Surface Coating via Thiol Anchoring. *J. Pept. Sci.* **2010**, *16*, 97-98. *31st European Peptide Symposium (31 EPS).* Sep 5-9, **2010**. Copenhagen (Denmark)

F. Rechenmacher, S. Neubauer, C. Mas-Moruno, P. M. Kleiner*, R. Burgkart, H. Kessler. Derivatization of Selective Integrin Ligands via Click-Chemistry for Surface Coating. *J. Pept. Sci.* **2010**, *16*, 97. *31st European Peptide Symposium (31 EPS).* Sep 5-9, **2010**. Copenhagen (Denmark)

F. Rechenmacher, S. Neubauer, C. Mas-Moruno, A. Bochen, P. M. Kleiner*, R. Burgkart, A. Schwede, J. P. Spatz, H. Kessler. Functionalization of Peptidomimetic Integrin Ligands with Thiols and Phosphonates for Surface Coating. *5th International Peptide Symposium.* Dec 4-9, **2010**. Kyoto (Japan)

S. Neubauer, F. Rechenmacher, C. Mas-Moruno, P. Kleiner*, R. Burgkart, J. Polleux, H. Kessler. Functionalization of Peptidomimetic Integrin Ligands with Thiols and Phosphonates for Surface Coating. *10th German Peptide Symposium.* Mar 7-10, **2011**. Berlin (Germany)

C. Mas-Moruno, P. Kleiner*, J. Beck, L. Doedens, S. Cosconati, L. Marinelli, R. Burgkart, H. Kessler. Cyclic RGD peptides: chemical tools to modify cell biology. *Biopolymers (Peptide Science)* **2011**, *96*, 482-483. *22nd American Peptide Symposium.* Jun 25-30, **2011**. San Diego (USA)

F. Rechenmacher, S. Neubauer, C. Mas-Moruno, P. Kleiner*, R. Burgkart, J. Polleux, H. Kessler. First coating of titanium with αvβ3 or α5β1 selective peptidomimetics: design, synthesis and biological evaluation. *Biopolymers (Peptide Science)* **2011**, *96*, 462. *22nd American Peptide Symposium.* Jun 25-30, **2011**. San Diego (USA)

S. Neubauer, F. Rechenmacher, C. Mas-Moruno, J. Polleux, P. Dorfner, H.-J. Wester, K. Pohle, A. Beer, H. Kessler. Biological Evaluation of the Distinct Integrin Subtypes α5β1 and αvβ3 via Surface Coating on Gold Dots and PET Imaging. *Biopolymers (Peptide Science)* **2011**, *96*, 461. *22nd American Peptide Symposium.* Jun 25-30, **2011**. San Diego (USA)

Oral communications and proceedings

P. Kleiner*, C. Horn, B. Laufer, M. López-Garcia, R. Gradinger, H. Kessler, R. Burgkart, Evaluation of the combined effect on structured and biofunctionalized implant surfaces for osteoblast adhesion, *Biomaterialien* **2008**, *9 (3)*, 127. *Jahrestagung der Deutschen Gesellschaft für Biomaterialen (DGBM).* Nov 20-21, **2008**. Hamburg (Germany)

P. Kleiner*, M. López-Garcia, F. Manzenrieder, J. Tübel, H. Kessler, R. Burgkart, Influence of different media to the adherence of osteoblasts, *Biomaterialien* **2009**, *10 (1)*, 51. *Internationale Biomechanik- und Biomaterialtage.* Jul 10-11. **2009**. München (Germany)

P. Kleiner*, C. Mas-Moruno, F. Manzenrieder, R. Burgkart and H. Kessler. How to investigate cell adhesion, IGSSE workshop on Functional Biosurfaces and Sensors. *Workshop Functional Biosurfaces & Sensors (IGSSE meeting).* Jun 28-30, **2009**. Unterach am Attersee. Salzkammergut (Austria)

C. Mas-Moruno, P. Kleiner*, F. Manzenrieder, R. Burgkart and H. Kessler. Integrin adhesion to titanium surfaces mediated by RGD-modified peptides. *Workshop Functional Biosurfaces & Sensors (IGSSE meeting).* Jun 28-30, **2009**. Unterach am Attersee. Salzkammergut (Austria)

P. Kleiner*, C. Mas-Moruno, F. Manzenrieder, R. Burgkart and H. Kessler. Investigation of osteoblast adhesion influenced by RGD-peptide coating and surface properties *4th IGSSE Forum.* Jun 14-18, **2010**. Raitenhaslach, Burghausen (Germany)

C. Mas-Moruno, P. Kleiner*, F. Manzenrieder, S. Neubauer, R. Burgkart, H. Kessler. Investigation of integrin and osteoblast adhesion influenced by RGD-peptide coating and surface properties. *5th International Peptide Symposium.* Dec 4-9, **2010**, Kyoto (Japan)

F. Rechenmacher, S. Neubauer, C. Mas-Moruno, P. Kleiner*, R. Burgkart, J. Polleux, H. Kessler. First coating of titanium with αvβ3 or α5β1selective peptidomimetics: design, synthesis and biological evaluation. *22nd American Peptide Symposium.* Jun 25-30, **2011**, San Diego (USA)

P. Dorfner, R. Burgkart. Overview of challenges in the development of implant materials. *BIT's 4th Annual World Congress of Regenerative Medicine & Stem Cell.* Nov 11-13, **2011**. Beijing (China)

* Petra Kleiner was my maiden name.

LIST OF ABBREVIATIONS

ACN	Acetonitrile
ADP	Adenosine diphosphate
Ahx	ε-Aminohexanoic acid
ALP	Alkaline Phosphatase
BCIP	5-Bromo-4-chloro-3-indolyl
BMU	Basic multicellular units
BSA	Bovine serum albumin
c	cyclo
CI	Collagen I
Cbfa1	Core binding factor alpha 1
Cile	Cilengitide
DMEM	Dulbeccos modified eagle Medium
DMSO	Dimethyl sulfoxide
ECM	Extracellular matrix
ELISA	Enzyme linked immunosorbent assay
ESB	European Society for Biomaterials
FA	Focal adhesions
FAK	Focal adhesion kinases
FCS	Fetal calf serum
FDA	Fluorescein diacetate
FN	Fibronectin
GFP	Green fluorescent protein
Gly	Glycine
HA	Hydroxyapatite
HFIB	Human fibroblasts
HRP	Horseradish peroxidase
IC	Inhibitory concentration

IgG	Immunoglobulin G
IgSF	Immunoglubulin superfamily
IHC	Immunohistochemical
IGSSE	International Graduate School of Science and Engineering
LDH	Lactate dehydrogenase
Lys	Lysine
MSC	Mesenchymal stem cells
OC	Osteocalcin
PBS	Phosphate buffered saline
PCR	Polymerase chain reaction
PLL-g-PEG	Poly-(L-lysine)-g-polyethylene glycol
PMMA	Poly(methylmethacrylate) (= bone cement)
RGD	Arginine-Glycine-Aspartic acid
RT	Room temperature
SAOS	Sarcoma osteogenic cells
SEM	Scanning electron microscope
TUM	Technische Universität München
UC	Uncoated

Motivation

In 2006 the Technische Universität München won the "elite" status in the Excellence Initiative by the German Federal and State Governments. The International Graduate School of Science and Engineering (IGSSE) as a part of the internal program "TUM – The entrepreneurial university" is an interdisciplinary scientific institution of the TUM. It aims to integrate natural and engineering sciences through graduate and postgraduate education rooted in a strong research foundation. Our research team was one of the first groups established within the IGSSE initiative. It consisted of scientists from the Department Chemie and physicians from the University Hospital Klinikum rechts der Isar. The project "Coated implantation surfaces" was launched with the aim to develop implant materials by using RGD peptides as surface coating molecules. The amino acid sequence RGD stands for arginine, glycine and aspartic acid. This sequence is known to display the integrin-binding site of many adhesive proteins and therefore boosts cellular adhesion and improves the biointegration of implants. This approach combines basic science and expertise of the Organic Chemists in the field of RGD peptides together with the more application-oriented and broad knowledge of Orthopedics from the Medical Sciences. Whenever a natural hip joint gets replaced by an artificial joint, besides the adoption of the biomechanical function, a fast and stable contact between the bone cells and the implant material is required. This process (termed osseointegration) still remains a major challenge in medicine. The examination of this topic from different scientific disciplines provides new perspectives, insights and approaches to improve osseointegration.

The objective of this project was:
- to build up an *in vitro*-system for studying integrin-related adhesion
- to investigate different components of the coating peptide
- to study different surface topographies and their influence on integrin binding
- to establish a cell system that ensures a stable osteoblast phenotype
- to examine the effect of RGD peptides on different surfaces with respect to osteoblast adhesion
- to improve our knowledge on different aspects in the adhesion process

We succeeded in achieving comprehensive results for each particular issue and described our findings in the following.

1 INTRODUCTION & BACKGROUND

We start with a retrospective of implant development from the first approaches to the state-of-the-art technology nowadays. Further we define important terms in the field of biomaterials. The function of osteoblasts, the adhesion process and the role of integrins are explained in detail and different approaches to measure cell adhesion are critically reviewed. In addition, the biofunctionalization of surfaces is discussed; we focus on the usage of RGD peptides, as these were highly promising molecules for optimizing cell adhesion.

1.1 Optimization of bone implants by improvement of osseointegration

In 1890 Themistocles Gluck became pioneer of the endoprosthetics, when the idea of an artificial implant became true with a knee implant consisting of ivory.[1] Around 30 years later (1923) Smith-Peterson attempted the substitution of a femoral head with a bowl of acrylic glass that failed due to material properties.[1,2]

Mechanical strong and corrosion resistant CoCrMo alloy (invented by Venable and Stuck) lead to fast improvements in this field, Moore and Thomsen for instance developed the shaft prosthesis in 1950.[3]

Beginning of the 1960ies Charnley et al. incorporated the challenge of tribology, meaning the friction between the artificial acetabulum and the head of the femoral part.[1] His low-friction principle included the choice of appropriate material and an optimized geometry of the slide face. Further, he was using polymethylmetacrylate (PMMA), also known as bone cement, for anchoring the implant, which was so far only applied in dentistry. These settings induced a broad spreading of total hip endoprosthetics.[2]

Besides the implantation with cement which was limited to the bone-cement-interface, another approach of finding an acceptable solution was the development of cement free ways for fixation. The two approaches differed in their point of origin for a joint substitution: In the passive anchoring with cement, the interface can be defined as bone-cement-various implant materials, like metal, ceramic, or PE. In this case the success is measured by means of no tissue loss or decrease. Otherwise the interface bases on the direct contact of the bone tissue with the implant material applied. This situation is termed as active anchoring and can result in a tissue proliferation and thus in osseointegration. Ring implanted in 1964 the first cement free metal/metal total hip prosthesis and also Mittelmeier forced the fixation without cement.[1,3,4]

The advantage of this approach is based on the proximal load transmission, which causes a more physiological bone load and therefore less stress shielding. Furthermore, in cement free anchoring a lower bone loss occurs during a revision.

In this regard many different ways of surface improvement were challenged ranging from surface area increase by simple mechanical treatment to more sophisticated methods like the usage of surface with tripods. In this case the implant surface was laced with defined elements (ranging from 0.65 mm to 3 mm) that lead to a sponge like metal surface.[2]

Within the next 20 years the developments were focusing on the anatomy of the proximal femur and in 1980 the first "GHE" hip pedicle was implanted (Figure 1).

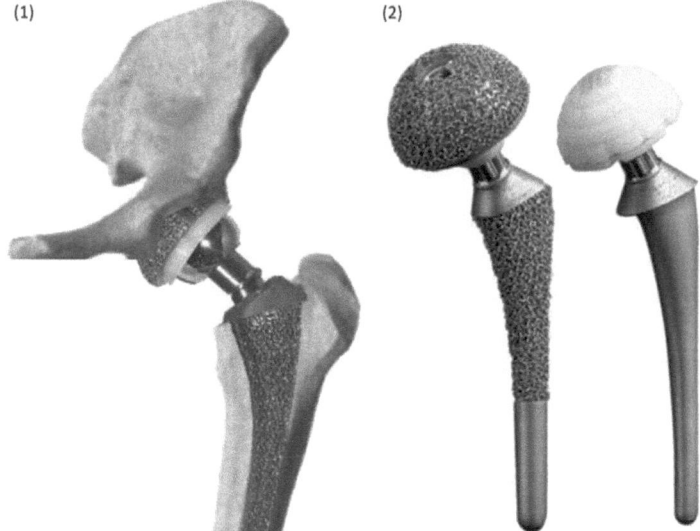

Figure 1: (1) Schematic of an implanted artificial hip joint with the remaining natural pelvis and femur (2) Femoral shaft type GHE. Left side: cement free version, right side: cemented version[5]

In this approach, the specially designed shaft takes the natural form of the proximal femur into account (the anatomic adapted form of this implant shaft is oval in the proximal area and comes up conical at the distal end) and leads, therefore, to a better outcome and less fractures.[5] Biomechanical considerations need to be the basis of an endoprosthesis, because the loading on the material in the joint is extremely high. The best possible load transmission is required aiming an uniformly and physiological load transfer between implant and bone, so relative movements between these should be avoided or minimized.[6]

Due to the fact that many terms have been used with different meanings, in 1986 the European Society for Biomaterials held a conference titled "Definitions in Biomaterials" to standardize the nomenclature and the following definitions were settled:[7]

- **Biomaterial:** A non-viable material, used in a medical device, intended to interact with biological systems
- **Implant:** Any medical device made from one or more materials that is intentionally placed within the body, either totally or partially buried beneath an epithelial surface
- **Biocompatibility:** The ability of a material to perform with an appropriate host response in a specific application;
  Two terms are distinguished:
  > (1) Compatibility of structure meaning the adaption of the form and structure of the implant material to the mechanical characteristic of the tissue
  > (2) Compatibility of surface meaning the adaption of the chemical, physical, biological and morphological surface properties of the implant material with the aim to obtain a good interaction with the accepting tissue[7]

Another important term is the osseointegration that describes the bony anchoring of bone tissue with the implant material. This term was introduced by Brånemark in the 1950ies, when he discovered that bone can integrate with titanium components.[8,9] The process of osseointegration requires a strong biochemical and mechanical interaction between the implant surface and the surrounding natural bone tissue. These interactions take place and are crucial within the first hours of adhesion and determine the stabilization and long-term success of the biointegration and consequently the implantation.

In this study we concentrated on the interface between implant materials and bone tissue, so, in the following, we focus on the cement free fixation of implants. In general an artificial hip joint consists of an articulating bearing (femoral head and cup) and a stem. With the cement free implantation the positioning of the articulating bearing is adapted to the natural components. So movements inside the hip joint remain quite physiologically (Figure 1). The cup is fixed by press-fit in the pelvis, meaning that only so much natural acetabulum is removed that the hole fits the design of the cup. For secure anchoring of the femoral head the stem is positioned in the intramedullary canal of the femur.

All practical implant materials need to fulfill the key parameters.[10,11] These requirements are mainly applicable for materials that are applied without cement, however, also material for cemented implants (here polyethylene is most common) demand the following criteria:

- Chemical composition that is biocompatible to prevent damage of surrounding tissue
- Mechanical strength to insure a lasting load transmission between the bone tissue and the implant material
- High resistance to corrosion for prevention of corrosive damage of implant within the body
- High wear resistance for decrease of wear debris

Cobalt-chromium-molybdenum and titanium with alloys are the most often used materials in the case of direct bone contact. CoCrMo (like CoCr28Mo6) or also CoNiCrMo possess high endurance strength, tension elastic limit, and low notch sensitivity. However, there are restrictions in potential allergization, as these materials release metal ions. Therefore, titanium grade II and IV, TIAl6V4 (which is termed grade V) and also TiAl5Nb7 are commonly used. Additional advantages of these materials are their good biocompatibility, the formation of connective tissue capsule, the low E-modulus that results in smaller stress shielding.[2,11-18]

The characteristic of titanium and its alloys is a thin oxide layer that is spontaneously formed on contact with air. This layer can be assumed as part of the titanium surfaces. Due to the physicochemical properties of this layer, the material is less inert for corrosion and shows a good biocompatibility, so proteins, e.g. albumin, laminin, fibronectin, fibrinogen, can adsorb from biological fluids.[11] After a non-specific protein adsorption, neutrophils and macrophages are the first cells on the interface to the implant.[11,19] We will focus on this issue in the next chapter.

Also titanium and its alloys, as the materials of choice, cannot meet all desired requirements. Therefore, more and more composite materials are developed and on their way to be clinically applied. Recently, the first ceramic-on-metal total artificial hip system was approved in the U.S. The Pinnacle CoMplete Acetabular Hip System by DePuy is the first to combine a ceramic ball and a metal socket.[20]

Besides implant criteria such as surface properties, form, or composition, also other parameters affect the process of bone ingrowth. For instance patient immanent factors, like bone quality, sex, or age influence the osseointegration as much as the operating experience of the surgeon.[21] Osseointegration is a complex process and some parts are still not understood in detail, but the achieved treatments lead to satisfying results. Over 90 % of the implants stay well functional for ten years after a hip replacement, 20 years after surgery the chance for an intact joint is about 80 %.[22,23] By 25 - 30 years after surgery, about 50 % of hip replacements are still working well. This data includes both cement free and cemented fixation.

However, different reasons are possible for the wear out of prosthesis. By far the most cause for a revision is the aseptic loosening (74 %).[22,24,25] In this case polyethylene (often used as cup inlay) is degraded and leads to a decline of mechanical and tribological feature, causing more and more wear debris per time interval, so the body's defenses are overstrained. In addition, the fitting of the implant is no longer accurate and therefore pain occurs (Figure 2).[23,26] Other reasons of another hip replacement include infection, breaking of the prosthesis, fracture of the bone around the prosthesis, and other complications. In most situations when pain occurs related to the artificial joint, the implant needs to be at least partially replaced.

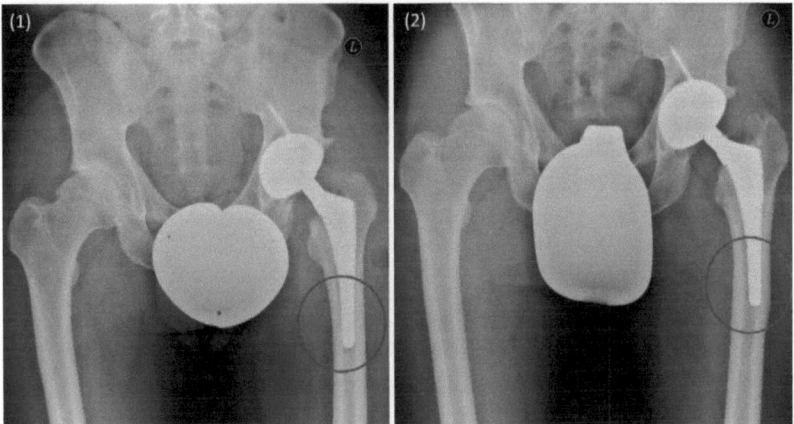

Figure 2: X-ray of a human pelvis with a total hip replacement at the left side (1) Shaft of the implant is in line with the femur and the force is directed vertically (2) After one year the fit of the implant is distracted, the shaft presses laterally on the bone.

Hip revisions are usually more complicated surgeries than the first replacement. The surgery is more extensive, because the old hip implant needs to be removed and the adequately secure fixation of the second implant is more complex. The quality of the bone becomes less stable, because the patients are older and less tolerant towards long surgical procedures. Taken these factors together the outcomes are not as good as for the first surgery.

In 2009 around 209,000 total hip implantations were performed in Germany.[27] The statistics over the last years demonstrate a constant increase of implanted hip joints. Figure 3 illustrates the increase of surgeries over the last 40 years in Sweden, where all orthopedic surgeries have been documented since 1979. According to this study the risks of a revision are slightly higher for a cement free implant then for one that is fixed with cement. Cement free prostheses are revised more often owing mainly to fractures close to the prosthesis and technical problems, while the most common cause of a revision from cemented prosthesis is loosening.[28] The major drawback of bone cement is the fact that the application of PMMA in the prepared cavity induces a high pressure within the bone. This fact extrudes adipoid cells from the bone into the vascular system and can lead to fat embolism which is often lethal.[29,30]

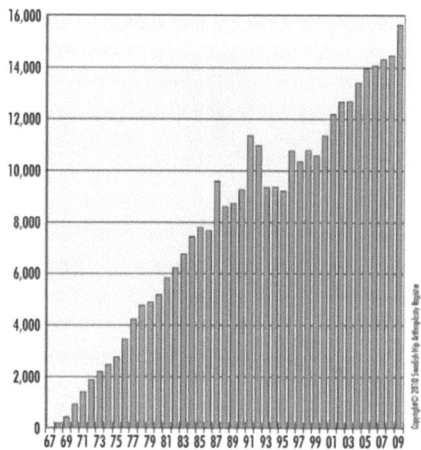

Figure 3: Primary total hip replacement in Sweden. Numbers of primary total hip arthroplasties performed in Sweden between 1967 (6 operations) and 2009 (15,646 operations), inclusive.[28] The population grew in that period by a factor of 1.2 from 7.868 to 9.2985 million.

In summary, more and more cement free prosthesis are implanted.[31-33] Since the implant has direct contact to the bone structure, a fast and successful osseointegration is essentially required and further improvements in this field are necessary. New successful orthopedic biomaterials must support adhesion, organization, differentiation and matrix mineralization of osteoblasts and osteoprogenitor cells.[34]

1.2 Osteoblast adhesion is a complex process

In this section an outline of the bone structure and the involved cells, e.g. the role of osteoblasts is given. Further osteoblast adhesion and the special role of integrins, the main adhesive receptors in the cell membrane, are highlighted and last, but not least, various methods to measure cell adhesion *in vitro* are specified and compared.

1.2.1 Bone structure and role of osteoblasts

Human bone consisting of different kind of cells undergoes a constant remodeling. This enables the bone structure to respond and adapt to mechanical stresses and loadings.[35] So called basic multicellular units (BMUs) control and manage the coupled process of bone resorption and replacement with newly built bone. This process regards around 20 % of the cancellous bone surface at any time. Osteoblasts and osteoclasts are mainly involved in this process, following an activation-resorption-formation sequence of events.[36]

The close association of inorganic minerals (e.g. hydroxyapatite) and organic macromolecules (collagen I, proteoglycan) features the strength of bony tissue.[37] This porous mineralized structure made up of cells, vessels, and calcium compound crystals varies strongly according to the type and region of bone. In the normal mature skeleton two types of bone structure can be observed: cortical and trabecular bone. These differ strongly macroscopically and microscopically, but are chemically identical.[38] On the one hand the dense and compact cortical bone, which comprises 80 % of the skeleton, is mainly calcified. It also has a slow turnover rate and its main function is to provide mechanical strength. Trabecular bone, on the other hand, exhibits a major metabolic function. Therefore, it has a higher remodeling rate and shows a lower density. Around 80 % of the bone surface is found inside the long bones.

Responsible for bone formation are osteoblasts derived from mesenchymal stem cells (MSC) (Figure 4). Osteoblast differentiation is a multistep process that is still not fully understood. The transcription factor Cbfa1/ Runx-2 plays a crucial role, when stem cells differentiate into osteoprogenitor cells and then to preosteoblasts. These cells mature finally to osteoblasts that lose their ability to divide, but produce unmineralized organic compounds of bone matrix. By mineralization of this osteoid, around 10 - 20 % of all osteoblasts become osteocytes, some develop to lining-cells, but 65 % undergo apoptosis after their performance.[34,39,40] Active osteoblasts are characterized with a cuboidal, flat morphology (20 μm in diameter), possess a large Golgi apparatus and show a large amount of rough endoplasmatic reticulum. As they are in charge for the osteoid calcification (hydroxyapatite), osteocalcin, collagen type I, alkaline phosphates, osteopontin, and osteonectin are typical markers for these kind of cells.[41-44]

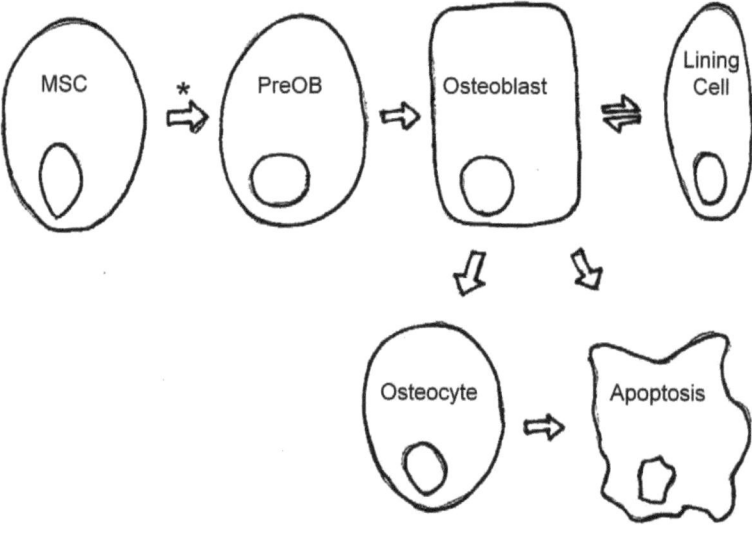

Figure 4: Schematic overview of differentiation of MSCs to osteoblastic cells. All steps are influenced by soluble factors in the media (growth factors) and also by the surface properties. *Several transition steps from mesenchymal stem cells via osteogenitor cell to PreOB.[45,46]

The classification of osteoblasts with respect to other cell types is challenging, because *in vitro* fibroblasts show a similar expression pattern. Ducy et al. even stated that osteoblasts can be viewed as a sophisticated fibroblast, because all the genes expressed in fibroblasts are also expressed in osteoblasts, and, vice versa.[41] Only two osteoblast specific transcripts have been identified to be only expressed in osteoblasts: (1) the one encoding for osteocalcin and (2) the other encoding for the

Core binding factor alpha 1(Cbfa1).[41] Further osteoblasts, as the bone builder, produce various ECM components and even in cell culture they are able to start the mineralization process. Therefore, osteoblasts can also be distinguished versus fibroblasts by alkaline phosphatase (ALP) production.

It is worth mentioning that the interaction of MSCs with surfaces, e.g. titanium, can support the differentiation into osteoblasts,[47] so detailed studying of the interference between stem cells, as well as bone cells and implant material promotes the development in this field. There is a high interest in differentiating stem cells into active osteoblasts. For the development of new therapeutic technologies for cell therapy, knowledge about the cellular and molecular events of osteogenic differentiation from MSCs is necessary. Not only local bone defects can be examined by side-directed delivery of MSCs in appropriate carrier vehicles, but also more general situations, like osteoporosis. By systemic administration of culture-expanded autogenic MSCs or through biopharmaceutical regimens based on the discovery of critical regulatory molecules in the differentiation process this disease may become treatable.[48] Currently autograft bone is used as the gold standard for these situations; however, there are major disadvantages, as only relatively small amounts of autograft are available and the harvest process is associated with significant morbidity.[49]

For *in vitro* studies the application of osteoblasts or bone-like cells is appropriate to obtain knowledge about the behavior of bone tissue on implant materials. For distinct research questions either the use of cell lines or primary cells is preferred. The advantage and disadvantages of both approaches must be weight up against each other and the decision should be made according to the underlying scientific question.

In our study we favor the use of human primary osteoblasts, as we have:

- the possibility to obtain these cells directly from human patients, meaning the origin of the cells is human. Further, also these donors underwent already a total hip replacement
- the facility to handle primary cells and characterize them carefully
- the knowledge to avoid dedifferentiation *in vitro* for a certain period of time

1.2.2 Osteoblast adhesion and the influence of integrins

In general the adhesion process is similar for all adherent cells and can be divided into three sections.[50,51] However, the processes can overlap. Figure 5 summarizes the adhesion process in these three steps.

(1) **Initial cell anchorage**
    First initial contact between cell and substrate, van-der-Waals-forces are involved and the cells resist only gentle sheer forces
(2) **Cell spreading**
    Cell starts to spread and flatten over the substrate and integrins get active, integrin clustering starts
(3) **Organization of the cytoskeleton and building of focal adhesions (FA)**
    Organization of actin in stress fibers, strong anchoring of the cell on the substrate, formation of FA with up to 50 transmembrane, membrane associated and cytoplasmic proteins involved[51,52]

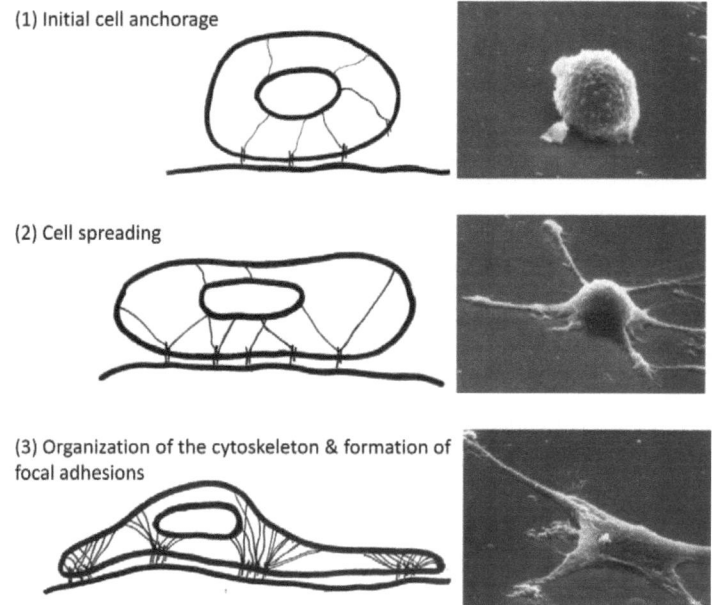

Figure 5: Sketch of the different steps within cell adhesion: (1) Initial cell anchorage (2) Cell spreading (3) Organization of the cytoskeleton and formation of focal adhesions. Images of cells from Drotleff et al.[53]

Four major classes of adhesion molecules on the cell membrane are responsible for the adhesion process. They belong to different families; however, these main classes can be defined:

- **Integrins** are transmembrane proteins and are responsible for the connection to other cells and the ECM and also for the communication among cells. A closer description of these most versatile type of adhesion receptors follows below
- **Selectins** composed of identical polypeptides chains, exhibit a lectin-like outer domain, which binds to oligosaccharide side chains of glycoproteins and are important for the inflammatory response
- The **immunoglobulin superfamily** (IgSF) includes cell adhesion molecules that are based on chains of immunoglobulin (IgG) and similar polypeptides. They include a large group of cell surface and soluble proteins that are involved in the recognition, binding, or adhesion processes of cells
- **Cadherins** (named for calcium-dependent adhesion) are a group of homophilic receptors that mediate tight mutual coupling of cells within cell monolayers. They play an important role in embryogenesis
- **CD44** is a cell-surface protein, which recognizes hyaluronic acid, a polysaccharide and is involved in cell-matrix interactions, cell adhesion and migration[54]

Integrins are the main receptors for binding extracellular matrix proteins.[55] They mediate and coordinate the anchoring process. Also the transmembrane signaling process is mainly directed by this kind of cell membrane receptors. These transmembrane receptors are non-covalently associated heterodimeric glycoproteins with a short intracellular C-terminal and a large extracellular N-terminal domain. In human 18 α- und 8 β-subunits were found to form at least 24 different integrin receptors.[56,57] Despite their wide variety, four main clusters are identified (Figure 6):

- **RGD receptors:** The RGD sequence found by Pierschbacher and Ruoslahti was demonstrated to be highly potent in boosting cell attachment.[58] The interplay of the RGD sequence in this group is identical for all integrin subtypes (α5, α8, αIIb, and αv with various β-subunits),[59,60] although the ligand affinity varies. This reflects the preciseness of interaction between ligand and specific α and β active side pockets[61]
- **Laminin receptors:** The sequences of laminin and collagens also contain the RGD motif; however, in this case, it is inaccessible for RGD specific integrins. The active side within the laminin has not been found so far, as the highly

selective integrins (α3, α6, α7, bound with β1 and α6β4) bind to different regions[61]
- **Collagen receptors:** Within the collagen-binding family, α subunits (α1, α2, α10, and α11 – all of these contain an αA-domain) are combined with β1. A glutamate within the collagenous GFOGER sequence provides the key cation-coordination residue.[62] The exact mechanism of the binding between the integrins and their ligand, collagen, remains unknown[61]
- **Leukocyte receptors:** The following integrins recognize related sequences in their ligands: α4β1, α4β7, α9β1, the four members of the β2-integrin, and αEβ7.[61] The binding motif is an acidic sequence LDV and the interaction takes place similarly to the RGD sequence at the junction between α and β subunits. Integrins β1 and β7 employ an aspartate residue for cation coordination, whereas glutamate is used in β2 ligands. The location of the binding site is another difference, for the β2 subfamily. It lies in an inserted A-domain in the α-subunit[61,63]

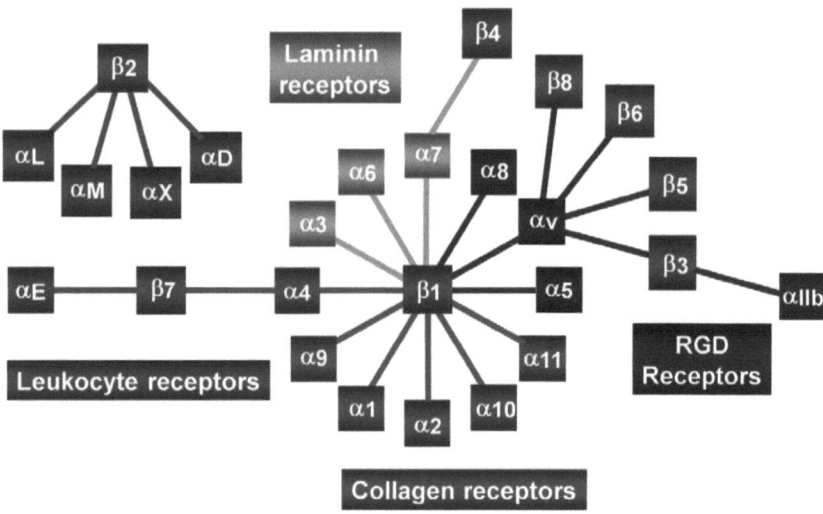

Figure 6: Overview of integrin subunits.[56]

It has been verified that the adhesion process depends upon integrin-mediated signal transduction and cytoskeletal molecules including vinculin, talin, actin filaments, and focal adhesion kinases (FAK).[64-67] As illustrated in Figure 7 the interaction between these molecules is complex and highly cross-linked. Actin filaments bind across talin, α-actin, vinculin and or the Arp2/3 complex to integrin receptors.[68] Vinculin, as one of the most connective molecules has binding sites for talin, and α-actin and secures the link from the cell membrane to the cytoskeleton.[66,67,69] Furthermore, Goldmann et al. stressed that both, talin and vinculin, are connected to integrin receptors to accomplish a double coverage of the sensitive adhesion process.[69] FAK is only expressed when integrin interacts with a ligand from the ECM.[70]

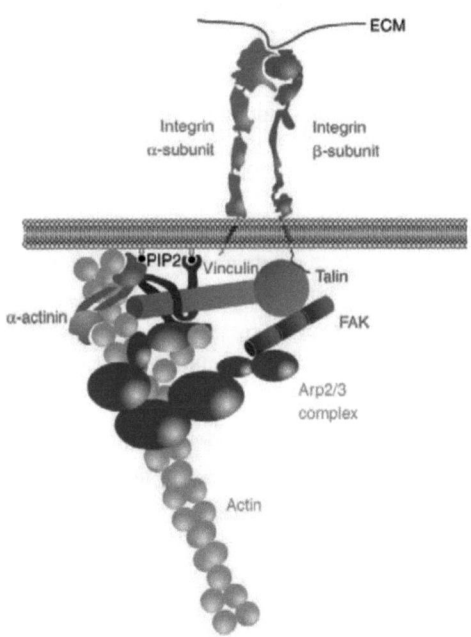

Figure 7: Overview of a focal adhesion contact illustrating how integrins connect the ECM with the intracellular cytoskeleton.[68]

The integrin family has the ability to transmit signals in both directions across the cell membrane, the so-called inside-out and outside-in signaling. For the inside-out signaling talin or vinculin need to bind on the cytoplasmic tail of the β-subunit, this binding results in a separation of the two subunits both at the transmembrane and the cytoplasmic domains of the integrin. The conformational changes of transmembrane domains propagate to the ligand binding headpiece and increase there the affinity of the integrin towards its ligand. Conversely, the outside-in signaling bases on ligand binding that induces the extrusion and thereby the same conformational changes on the α- and β-subunits on both sides (intra- and extracellular).[24,71,72] Zhu et al. showed that the extension is responsible and required for ligand binding during integrin inside-out-signaling and not a deadbolt regulates integrin activation.[72] In X-ray crystallography analysis a "knee" region between the globular head of the integrin receptor and the rod-like tails was found.[34] This flexible part is responsible for integrin activation, as there the conformational change cause the extension.[34] Some integrins require an additional binding site (synergy site) on the ligand for an optimal function.[73] E.g. for the full binding performance of α5β1 also the synergy site PHSRN besides the RGD binding site needs to be bound.[34,74]

The possible cross-talk between integrins even enhances the complexity of the system. For example Simon et al. revealed the regulation of α5β1-mediated cell migration towards fibronectin by αvβ3.[75]

After ligand binding integrins start to cluster together and build focal adhesion contacts. The signaling pathway within the cell is complex and involves the accumulation of several proteins. First FAK is autophosphorylated and the phosphorylation of tyrosine is followed by the recruitment of other proteins.[76] Overall the interaction of integrin and ligand affects the signal transduction and the expression of genes within the cell.

Integrins as the main adhesive membrane receptors have special tasks in bone metabolism and a defect in this system leads to pathologies as osteoporosis.[77] Diet-related lack of calcium or the menopause of women entails a lack of estrogen, stimulates osteoclast activity and therefore the resorption of bone. By the adhesion of osteoclasts the proliferation of osteoblasts is encouraged to antagonize the bone resorption, however, the activated osteoclasts resorb more bone than osteoblasts are able to build up. It also needs to be mentioned, that the integrin pattern varies strongly over the cell cycle.[78] In bone metabolism a narrowed osteoblast adhesion by inhibited integrins leads to a higher level of apoptosis and thus to a lower bone formation.[79]

Osteoblasts express various integrin receptors on their cell membrane. In literature the results are inconsistent due to different cell sources, media conditions, differentiation states, or applied staining techniques.[73,78] Several groups detected the following integrin subtypes on the membrane of bone cells: αv, α1-5, α5, α6, and β1, β3, and β5.[34,43,73,80] As heterodimers α1β1, α2β1, α3β1, α5β1 and αvβ3 were assured.[81,82] Shekaran et al. stated the β1 subunit to be the most dominant and highly expressed integrin in osteoblasts and additionally they found αvβ3 to influence osteoblast proliferation and differentiation in a negative way.[34] This statement about the properties of αvβ3 is contradictory, as other publications show that αvβ3-selective ligands promote osseointegration *in vitro*[83-85] and *in vivo*[86]. Additionally αvβ3 raised attention because of its important role in pathologies as ocular diseases,[87] acute renal failure,[88] and metastasis formation.[89]

1.2.3 Different approaches to measure cell adhesion

In literature many different ways of measuring cell adhesion are described. Due to the fact of interdisciplinary there are various approaches to detect the interaction of cells with artificial materials. Physicists developed distinct imaging techniques to investigate the adhesion process. The analysis of diverse metabolites is common by biochemists, whereas the engineers tend to fabricate a tool that measures the adhesion force mechanically.

Figure 8 demonstrates different methods for measuring cell adhesion that are found in literature. One can distinguish between various tests for metabolite detection, different kinds of cell morphology visualization, and the detection of sheer forces.

The first type of adhesion tests, namely the **determination of cell metabolites**, is based on the detection of adhered cells.[76,90-98] Usually the amount of metabolite is correlated to the cell number. The more cells remain on the surface, the higher is the assumed adhesion. The big advantages on this approach are the fact that cells are not manipulated during their adhesion process and the easy utilization. With these tests an average result over many different cells is obtained. However, there are also disadvantages, as it is almost impossible to track the adhesion process by end point measurements and the quantification of cell numbers is no direct adhesion strength in terms of force. Anselme et al. introduced the *Anselme adhesion assay*, as a model for calculating the defined adhesion force of the cells. Thereby, cells are treated with trypsin for different time periods and detached cells as well as adhered cells are analyzed. The results of the experiments together with different roughness parameters of the applied surface are summarized in a mathematical model to determine the adhesion strength. The overall aim is to predict the performance of various surfaces with distinct parameters in terms of osseointegration in advance.[91,92,96-98]

A second approach is the investigation of the **cell morphology**. The shape of cells that changes over time gives insights on the adhesion process.[78,90,95,99-101] All of these methods are performed either with or without staining and convince by the obvious visual proof when looking through the microscope. Additionally, the progress of the adhesion process can be visualized and analyzed over a longer period of time. However, in this case also no direct adhesion force is detected and one must pay attention to the potential interference between adhered cells and the applied staining protocol. Another disadvantage is the fact that usually only very few cells can be analyzed at once.

A third approach is based on studying the **sheer forces**. Various apparatus have been developed to detect and analyze the adhesion strength.[78,102-105] According to the principle of the particular system, adhered cells are scratched, spun or drawn off the surface, while detecting the strength needed. The obtained force is given in Newton and very suitable for comparison between different set-ups. However the artificial and non-natural interaction with the cells is the major drawback of all these methods. Cells are damaged and the set-up bases on single-cell measurements in strongly non-physiological surroundings.

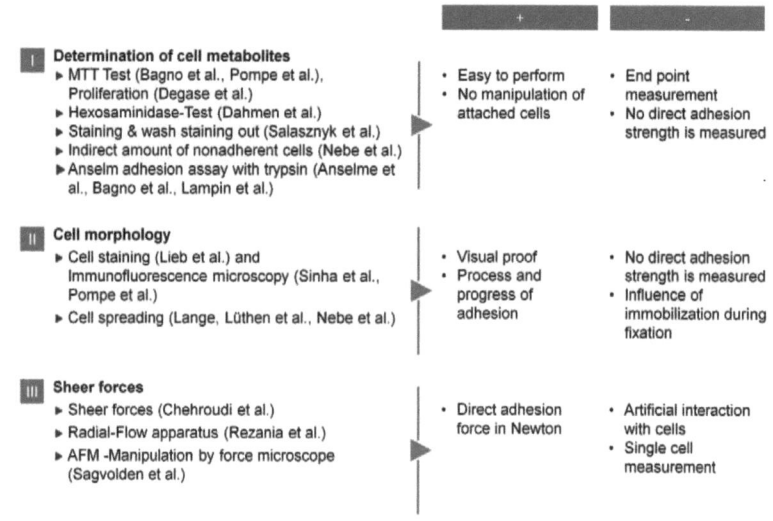

Figure 8: Overview of different approaches to measure cell adhesion.

In this study we performed the following cell adhesion tests; the choice of the assay was depending on the investigated question:

- **Hexosaminidase test** (Determination of metabolites)
  The amount of the enzyme hexosaminidase was detected and used to analyze the number of cells that adhered on different surfaces
- **Fluorescent cell staining** on Ti6Al4V disks (Cell morphology)
  By staining the cells with fluorescein diacetate (FDA) we investigated the level of adhesion after short time periods
- Tracking the **cell spreading** in an **optical microscope** (Cell morphology)
  The behavior of cells was tracked over a distinct time period to get an overview over the whole adhesion process
- Imaging the **cell spreading** with **scanning electron microscope (SEM)** (Cell morphology)
  Due to the resolution limitations of optimal microscopes adherent cells were also imaged by SEM to gain a better insight at cell morphology on different surfaces

1.3 Biofunctionalization with RGD peptides is a promising method for optimizing the selective adhesion

The biofunctionalization of implant materials is the topic of this chapter. Aiming rapid and specific cell colonization, mimicking biological surroundings by a natural ligand is a promising strategy in implant technology. The advantage of biofunctionalization lies in the control and regulation of specific interactions related on the defined structure of biomimetic materials. The functionalization of surfaces with bioactive molecules is performed in order to enable specific signaling and achieve the desired cellular response.[106,107] In this section we review the different approaches based on various coating molecules.

The RGD sequence is a common cell-recognition motif which is a part of integrin binding ligand, like fibronectin, fibrinogen, and vitronectin, von-Willebrand-factor as well as laminin.[108] As described in section 1.2.2, this sequence serves as the most effective and often employed sequence to stimulate cell adhesion on synthetic surfaces.[51] Further, it was used as a basis for the development of different integrin antagonists for cancer treatment.[109]

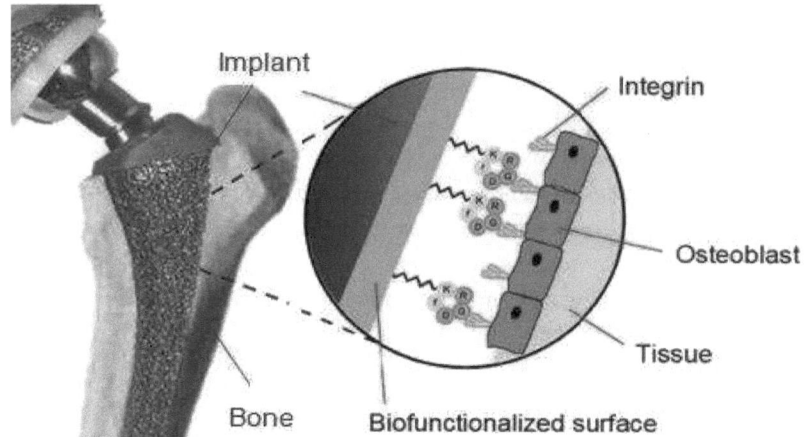

Figure 9: Sketch of the basic principle of surface functionalization on the orthopedic implant material.

The basic principle of surface functionalization on orthopedic implants with RGD peptides is illustrated in Figure 9. The peptide is immobilized on the implant material and by its structure it presents the RGD sequence towards the integrins on the cell membrane. The aim of this biofunctionalization is to provide a more attractive environment for osteoblasts and less capsulation with fibrous tissue for a better and faster osseointegration.[110,111]

López-García and Kessler described the evolution of different coating molecules containing the RGD sequence form the first approaches of extracting the relevant part from the whole ECM protein to recent developments of highly specific and selective peptidomimetics.[51] In the beginning the natural ECM proteins that contained the RGD sequence (so called 1st generation of adhesion stimulating molecules) were applied on the surface of implant materials.[112] These molecules benefit from their analogy to the natural adhesion protein. However, as they are enzymatic instable, they bear the risk for infections and inflammation and are immunogenic and difficult to anchor on the desired surface. With the 2nd generation of these molecules most disadvantages are overcome by reducing the protein to small synthetic RGD peptides,[113] so the risk of contamination and immunogenicity is eliminated. Additionally, the peptides possess a higher temperature and pH stability and can be packed with an increased density. However, these linear peptides show no selectivity on distinct integrin receptors and are still enzymatic instable. To eliminate these drawbacks the 3rd generation of peptides was developed. The structure and the conformation of the RGD sequence are crucial for its function and its stability. Therefore, the cyclization of this motif induces conformational stability leading to higher enzymatic stability. Furthermore, the specific selectivity for the ligand is highly influenced by the amino acid flanking the RGD sequence.[51,57,108,114-119] In addition, a preferred three-dimensional structure is obtained in order to achieve a better interaction with the specific integrin receptor. Synthetic peptides mimic the natural ligand in a way similar to the biological function of their considerably larger parental molecule.[50]

The structure of RGD peptides commonly used for coating different surfaces is illustrated in Figure 10. The molecule presents the specific and highly selective RGD sequence. In our study we used the sequence c(RGDfK) that has been shown to be selective for $\alpha\beta v3$ and $\alpha5\beta1$.[51,108,117,120]

The function of the spacer is to ensure an optimal presentation distance of the RGD sequence on the used surface. For the spacer unit different kind of molecules can be used, e.g. aminohexanoic acid, derivates of polyethyleneglycol, and also photoswitchable and photolabile units have been tested.[121] A minimal interspace of 3.5 nm was reported for an effective integrin-mediated cell binding.[51,57,110,122] The bottom line is that integrins on the cell membrane must be able to interact with the adhesion sequence of the peptides coated on the surface.[50]

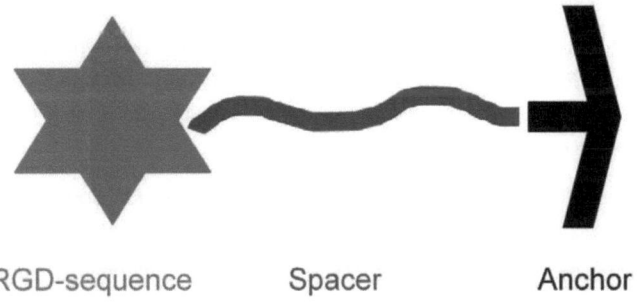

RGD-sequence   Spacer   Anchor

Figure 10: Structure of molecules with RGD sequence, spacer and anchor unit used for coating the implant material.

To ensure a stable and lasting interaction with the given material, the peptides need a suitable anchoring unit. As many different materials have been studied, a series of anchors was established, suitable for the particular surface. For example, a direct immobilization on the oxide layer of titanium or its alloys can be achieved by using either phosphonic acids[123-125] or thiol functionalities.[126,127] Although titanium surfaces have been reported to be attractive for thiols,[128] this anchoring group can be easily oxidized with concomitant loss of attachment.[129] Auernheimer et al. have demonstrated that the tetraphosphonate anchor provides a highly stable binding to Ti6Al4V under different working conditions (e.g. low pH) and that it is possible to sterilize the coated surface without losing osteoblast adhesion properties.[130] A corresponding study for the thiol-anchored peptides remains yet to be performed. It should be also mentioned, that the production of the phosphonate anchoring groups is synthetically more demanding than that of other units. Other materials like PMMA can be coated by an acryl amide anchor[122] and thiol units stick also on gold surfaces.[131]

Less relevant for clinical applications, but important for scientific investigations, are molecules like BSA or streptavidin, which have also been functionalized with RGD peptides.[51,122] A tool to reduce protein binding and adsorption from serum to the surface was established by the use of poly(L-lysine)-g-polyethylene glycol (PLL-g-PEG). PLL-g-PEG form a combined structure with positively charged primary amine groups of the PLL bound to the negatively charged metal oxide surface, while the hydrophilic and uncharged PEG side chains are exposed to the solution phase. Copolymer architecture is an important factor in the resulting protein resistance.[132] Therefore, the influence of peptides immobilized on side chains of the PEG brushes can be studied in an artificial surrounding aiming for a better understanding of the peptide-cell-interaction.

The use of conformationally constrained cyclic RGD peptides, which are highly active and selective for integrins expressed by osteoblasts (e.g. $\alpha v \beta 3$ and $\alpha v \beta 5$), has been shown to efficiently enhance the adhesion of osteoblasts *in vitro*[110,122,130,133] and improve the *in vivo* bone growth[31,110,125,126,130].

Publications differ regarding RGD peptide used, surfaces functionalized, cell types or animal models investigated and so on. Although it is therefore difficult to directly compare results in distinct publications, it is worth to mention some findings.

Kantlehner et al. reported an increased proliferation of osteoblasts (from different origin, like rat, mouse, human) on c(RGDfK) coated PMMA surfaces after 22 days. In their rabbit model the researchers showed an induction of enhanced and accelerated cancellous bone ingrowth and direct contact areas between the bone and the implant material when RGD coated, whereas the uncoated implants were surrounded by a fibrous tissue layer.[122] Elmengaard and coworkers investigated the effect of RGD coating on plasma-sprayed Ti6Al4V implants by histomorphology and push-out tests.

The cylindrical implants were inserted for four weeks in the proximal tibia of mongrel dogs. They found a two-fold increase of bone ingrowth for RGD coated implants and a reduced fibrous tissue ingrowth. Furthermore, a higher bone volume was detected within the 0 -100 μm zone and an enhancement in shear stiffness and energy to failure was observed.[31] RGD coated smooth titanium implants were implanted intramedullary in rat femora by Ferries et al. Although no differences in the mechanical fixation or implant bone coverage were found, the researchers detected increased bone thickness around peptide-coated implants.[86] Schuler et al. also investigated the effect of the RGD sequence on PLL-g-PEG functionalized titanium surfaces with different cell types, like epithelial, fibroblasts, osteoblasts. They observed more osteoblasts on rough and more fibroblasts on smooth surfaces, but all cells preferred bioactive substrates containing RGD. However, no synergetic effect of RGD peptide coating and surface topography was found.[134]

Another crucial parameter for the adhesion is the peptide concentration and therefore the peptide density on the surface. This issue was intensively investigated by many groups.[134-136]

Healy and Rezania studied the density of RGD peptides needed on the surface to achieve a specific cell response. For example, they stated that a peptide density of 0.6 pmol/ cm^2 promotes initial cell adhesion and calcification of the synthesized extracellular matrix.[136,137] They also found in their study that raising the concentration up to 3.8 pmol/ cm^2 did not have any further effect.[137] These findings are based on the assumptions that a spread cell has a contact area of 50 $μm^2$ and contains ~10^6 receptors that recognize the RGD sequence. Per 1 $μm^2$ 2000 receptors are present and surfaces coated with RGD peptide concentrations above 0.6 pmol/ cm^2 result in immobilized ligand densities of ≥ 3700 molecules/ $μm^2$. Therefore, this concentration should already saturate all of these surface receptors.[136] Ward et al. and Mooney et al. observed similar effects.[138,139]

The function behind the molecular arrangement of single integrin on the cell adhesion process was investigated by Arnold et al. They used RGD peptide coated gold nanodots with a diameter below 8 nm. This size allowed only the binding of one integrin per dot. By testing surfaces with various arrangements of these dots they showed that a maximum spacing between 58 and 73 nm is necessary to support cell adhesion and focal adhesion.[135]

As osteoblasts need to adhere and spread, but also to differentiate, Siebers hypothesized that the ideal bone-contacting implant should not be covered with too many peptides.[73]

Another aspect for the interaction between peptide and material was analyzed by Okamoto and Matsuura. They suggested that RGD peptides contribute to the osteoconductive effect of hydroxyapatite more than titanium, meaning that hydroxyapatite attracts more RGD containing ECM proteins than titanium.[140,141]

Besides the intense studies on the RGD sequence, the sequence located in heparan sulfate proteoglycan presents another approach for the enhancement of osteoblast adhesion. This osteoblast-specific mechanism is based on various identified heparin-binding domains.[142] Bagno et al. investigated a sequence mapped in the human vitronectin protein, namely (351-359)HVP.[91] Their experimental data revealed different bioactivity levels depending on the surfaces and peptides coated, but (351-359)HVP coating showed a similar adhesion capacity when compared to RGD grafting. However, Sawyer et al. compared the influence of peptide coated surfaces to serum-coated ones on the MSC attachment and spreading. They did not state any significant enhancement on hydroxyapatite (HA) neither for coating with RGD, nor proteoglycan-binding peptides.[143]

Furthermore, sequences isolated from other ECM proteins have been studied and proven to possess also positive effects on bone repair and orthopedic implant integration. For example, the α2β1 selective sequence GFOGER presents a collagen-mimetic peptide and improved in both *in vitro* and *in vivo* studies the osteoblastic differentiation and mineral deposition as well as bone regeneration and osseointegration.[111]

In summary, it can be stated that the RGD sequence had been studied extensively and many findings confirm the positive effect of RGD functionalized surfaces with regard to cell adhesion. The ongoing development in this field will lead to highly specific and selective peptidomimetics providing a promising strategy to improve cell adhesion addressing different scientific and medical questions.

2 RESULTS & DISCUSSION

In this work we focused on four main topics with respect to osteoblast adhesion. These subjects are schematically illustrated in Figure 11. The corresponding chapters will give answers to the key questions of each topic.

(1) **Primary osteoblasts are obtained from human bone**
How to obtain human primary osteoblasts? What are the characteristics of this cell type? How to culture osteoblasts avoiding dedifferentiation?

(2) **Investigated surfaces vary in terms of roughness**
How can Ti6Al4V surfaces be characterized? What are the relevant parameters for predicting cell adhesion?

(3) **Binding assays with integrins verify affinity of RGD peptides**
How does the structure (anchor/ spacer) of RGD peptides influence the integrin binding on different surfaces? What testing system suites best to investigate this binding?

(4) **Adhesion assays with cells show beneficial impact of RGD peptides at the early adhesion process**
How do osteoblasts adhere on RGD coated Ti6Al4V surfaces? Which RGD peptide type and surface structure shows the best results? What is the most relevant timescale for the adhesion? What conclusions can we draw form the *in vitro* assays?

In addition, we were able to perform an animal study in cooperation with CeramTec. Within the project "direct to bone" ceramic probes got implanted in the femora and tibiae of sheep. Irrespective of the different material, this study delivered insight into the performance of implants coated with RGD peptides *in situ*.

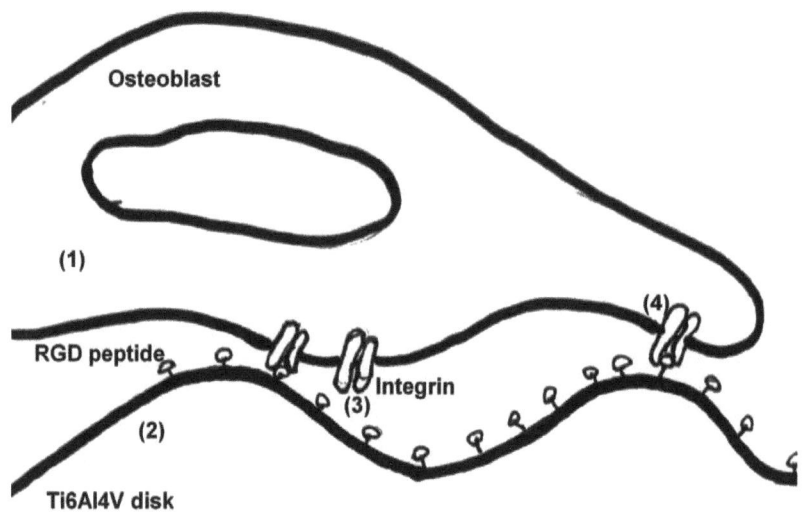

Figure 11: Overview of investigated topics: (1) Characterization of osteoblasts, (2) Characterization of Ti6Al4V surface, (3) Interaction of integrin and RGD peptide (Assays with integrins), (4) Interaction of osteoblast and RGD peptide (Assays with osteoblasts).

2.1 Primary osteoblasts were obtained from human bone

Experiments using primary human osteoblasts have several advantages compared to working with cell lines. As cancerous and immortalized osteoblastic cells have lost many of their osteogenic and adhesive features,[73,144,145] it is clear that primary osteoblasts are preferred to study cell-surface interaction.[146] We are working with bone cells of human origin. In the following we will use the term osteoblasts, although the cell type is a pre-stage of mature osteoblasts. Mature osteoblasts are not able to divide anymore and produce organic compounds of the bone matrix for mineralization. Therefore, mature osteoblast cannot be cultivated in cell culture.

The pre-osteoblasts, however, serves as an adequate cell model in the field of implant research. The orthopedic implants are directly exposed to this kind of cell population *in situ*. In general a suitable cell model is very important for scientific research dealing with medical devices, as it gives first useable insights for the acceptance and reaction of the tested samples on cellular level. It is also more cost efficient and less labor intense than *in vivo* models besides a better ethical tenability.

In this chapter we explain how we obtained and cultured primary human osteoblast. As primary cells tend to dedifferentiate while being cultured, the testing of specific cell-metabolites is crucial to avoid working with differentiated cells. In order to do so, we stained our cells for the typical osteoblastic metabolites, alkaline phosphatase (ALP),[42] fibronectin (FN),[43] type I collagen (CI),[43] and osteocalcin (OC).[41,44] Further the ratio of osteoblasts and fibroblasts was determined and we proved the osteoblasts to express αvβ3 on their cell membrane.

2.1.1 Extraction and culturing of human primary osteoblasts

During a total hip joint replacement surgery, cancellous bone was obtained from surgical waste and used to extract primary osteoblasts. Incubating the bone chips in low Ca^{2+} alpha media leads to a sprouting of osteoblasts. As illustrated in Figure 12, cells expand from the bone chip and adhere with the typical osteoblast morphology on the surface of the tissue culturing Petri dish. With the optimal culturing conditions osteoblasts grow autonomously onto the surface. This behavior is mostly depending on the media and its additives. Further, it must be mentioned, that primary cells can only be cultured for a distinct time period of some weeks and the number of passage is limited.[146]

To obtain a pool of primary osteoblasts, cells from six different patients were assembled in equally parts. We have selected male (3) and female (3), as well as young and old donors (18 - 63 years; see Table 1 for detailed information).

Table 1: Detailed information of donors for cell pool.

Internal number of donor	Age	Sex
1	18	Male
2	42	Male
3	44	Male
4	52	Female
5	39	Female
6	63	Female

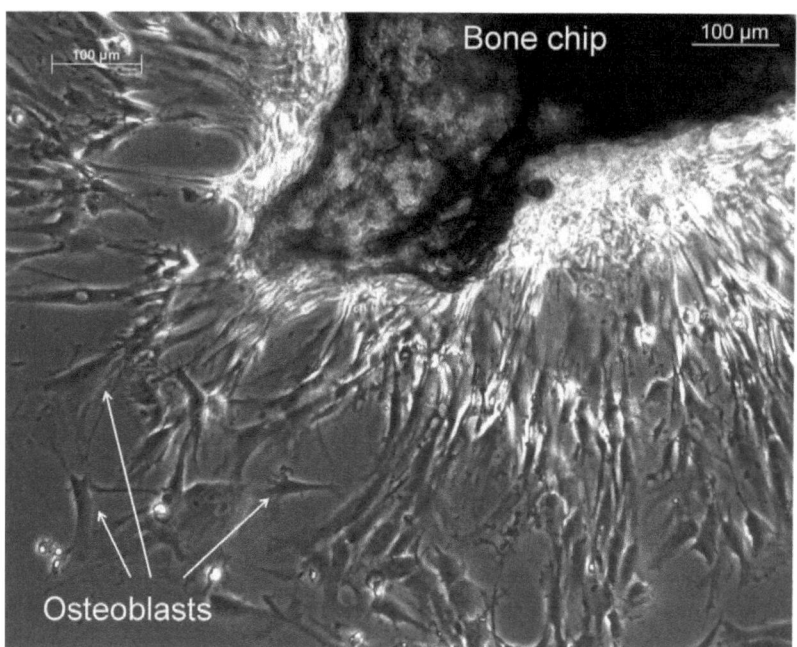

Figure 12: Sprouting of bone cells from bone chips, osteoblasts expand to Petri dish autonomously.

2.1.2 Qualitative characterization of osteoblasts

Testing the expression of cell-typical metabolites to ensure working with non-dedifferentiated osteoblasts is relevant, as we want to investigate the behavior of this cell type with respect to the implant material. It is important to prove that the pooled cells express characteristic bone markers.[41-44,147,148] Some cells will differentiate to fibroblasts or other cell types, but we demonstrated by the positive staining in Figure 13 that our cells express the common osteoblast-specific markers: ALP, FN, OC, and CI.

In cell culture osteoblasts and fibroblasts are very hard to distinguish by their morphology. Genetically both cell types are almost identically and osteoblasts can be viewed as sophisticated fibroblasts.[41] Even though fibroblasts secrete most of the extracellular matrix proteins, however, to a lesser extent than osteoblasts, only osteoblasts are responsible for building bone.

The bone specific ALP is a glycoprotein found on the surface of osteoblasts that has been shown to be a biochemical indicator for bone turnover.[149] Cells were positively stained for this hydrolase enzyme. The ratio of ALP positive and negative cells is around 75 %.[150] The presence of FN was affirmatively confirmed. This ECM glycoprotein is important for cell migration and adhesion. In a common cell culture the percentage of positive stained cells is higher than 90 %.[150]

Another important substance for testing the osteoblastic phenotype is OC, the most abundant non-collagenous protein in bone. The cells were stained positively for this metabolite (in general around 80 %[150]). OC is also known as bone γ-carboxylglutamic acid-containing protein, which binds to hydroxyl apatite and calcium. In addition, we tested the cells for Cl, a fibrous protein that is responsible for the strength and flexibility of the bone. Around 85 % of the cells are tested positively in an usual osteoblast culture of primary cells.[150]

These findings demonstrated that *in vitro* osteoblasts produce the molecules they synthesize in their original tissue, namely bone; thus they are able to induce their own *in vitro* mineralization.[151]

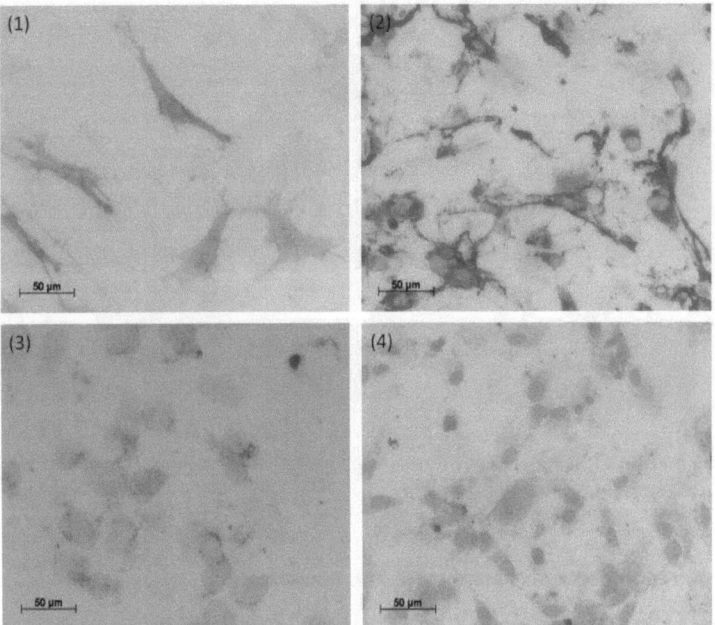

Figure 13: Detection of cell metabolites by immunocytochemistry. Human osteoblasts were stained for ALP (1), FN (2), Cl (3), and OC (4).

A nodule with surrounding osteoblasts is shown in Figure 14. The staining of nodules was performed to investigate the ECM building of the osteoblasts, which is the unique characteristic of bone cells and therefore a confirmation for culturing osteoblasts. The dark area within the nodule is due to a high accumulation of calcium. This natural behavior of the cells is objectionable for cell cultures, because cells agglomerate tight together and cannot be separated from each other anymore. The process of nodule formation was investigated in detail by Owen et al. and can be separated in three states of development: Proliferation, building and maturing of the ECM, and mineralization.[152] Nodules were proven to show characteristics of embryonic bone.

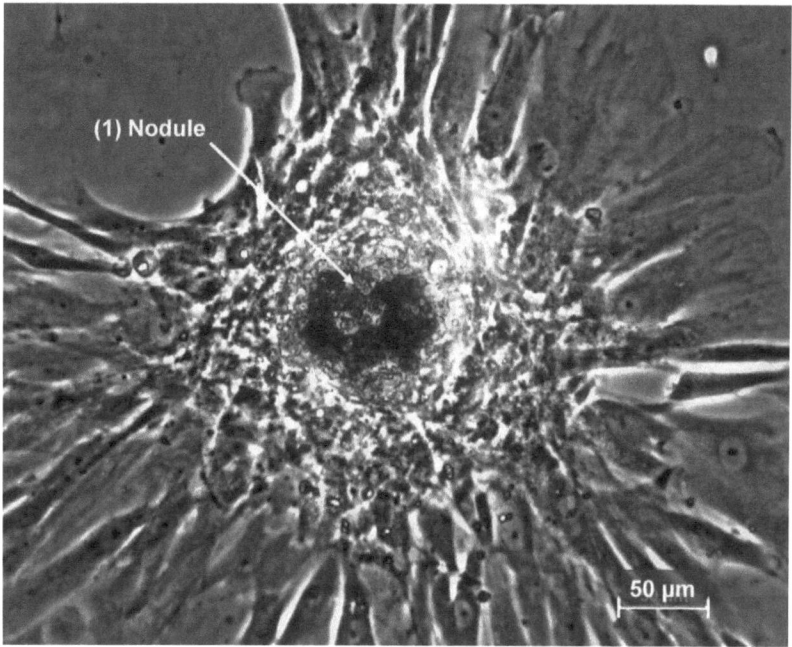

Figure 14: Image of nodule formation within an osteoblast culture. In α media osteoblasts generate nodules (1) autonomously through accumulation of calcium. This can be inhibited by the use of calcium-free media.

Determination between osteoblasts and fibroblasts

In order to characterize the ALP activity human osteoblast, sarcoma osteogenic cells (SAOS) and fibroblast cell cultures were stained for the phosphatase. The staining of

ALP with the substrate 5-Bromo-4-chloro-3-indolyl phosphonate (BCIP) is straightforward, cost-efficient and very fast.

The ratio of ALP positive cells to the total cell number is shown in Figure 15. SAOS cells are 99 % ALP positive; this is expected for that cell line. Within the pooled cells that were used for the experiments, 60 % of the cells are ALP positive and therefore osteoblasts. Within the single donors individual changes can be found, the amount of osteoblasts is ranging from 32 to 71 %. These numbers are in line with the findings of Robey et al.[148] The other cells present within our cultures are fibroblasts and the fact of having a mixture of cell is a usual characteristic, when working with human primary cell cultures.[148] Cells interact with each other and can stimulate the dedifferentiation process by the secretion of various factors. The low amount of ALP in donor 18, m can be explained by a genetically disposition, as in this age a total hip replacement is unusual. It is possible that the patient obtained special drugs in advance of the surgery. Due to data security it is not possible to investigate this closer; however we intended to obtain a cell pool with a broad mixture of different donors.

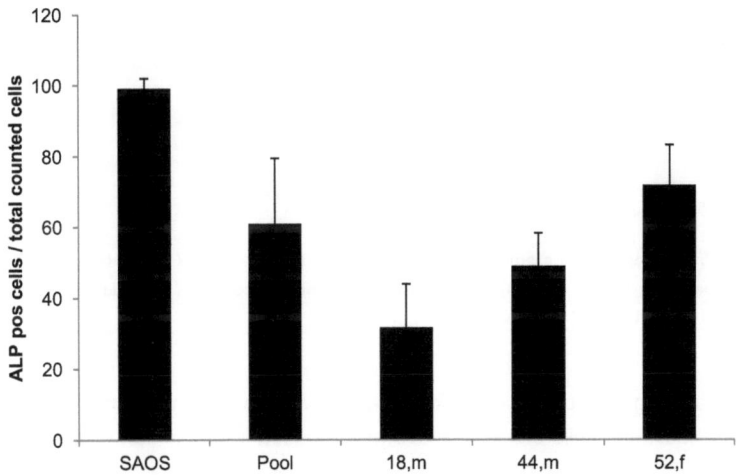

Figure 15: Ratio of ALP positive cells and total cell number in per cent. The SAOS cell line as well as the pooled human primary cells (Pool) and three different donors (18, m; 44, m; and 52, f) were investigated.

Figure 16 illustrates the staining results for primary human osteoblasts and the SAOS cell line. Not all cells of the donor 44, m are stained positive. ALP negative cells can only be observed in phase contrast modus and are unseen when observed in an optical microscope without phase contrast (compare (1) and (2)). In contrast to this, all SAOS cells remain visible in both settings and present thus a 100 % ALP activity.

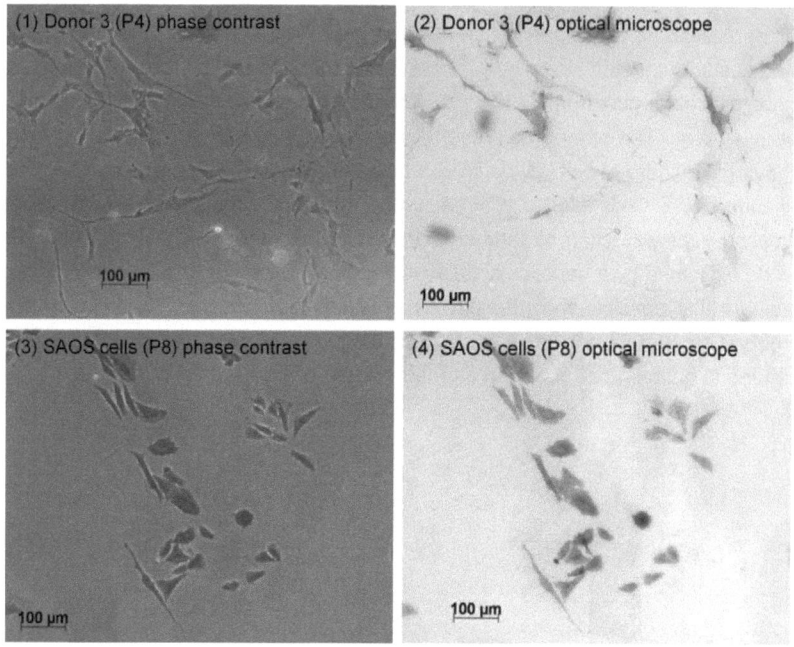

Figure 16: Images of ALP stained osteoblasts (1) and (2) donor 3 (44, m, passage 4), (3) and (4) SAOS (passage 8); (1) and (3) Cells observed in phase contrast, (2) and (4) Stained cells observed in light optical microscope.

When culturing cells over a distinct time a decrease in staining intensity is observed. At later passages cells are less active and produce lower amounts of ALP. This is displayed in Figure 17, the staining for ALP hardly appears in passage 9 compared to passage 4. Although it is possible to culture human osteoblasts even up to 15 passages without changes in the metabolite expression, more and more cells start to dedifferentiate over time. Primary and first-passage cultures have been maintained in low Ca^{2+} medium for periods up to 4 month.[148]

Another challenge in working with primary cells is the reproducibility of experiments. In contrast to cell lines, which have been studied for a long time, primary cells of each donor behave slightly different and make it difficult to reproduce and compare results over time. To ensure working with dedifferentiated cells and secure reproducibility, all experiments in this study were performed with cells below passage 6.

Also the method of culturing influences the cells, as Marques da Silva et al. showed in their study with static and dynamic culturing conditions.[153] In our experiments we maintained a classical static culturing system to reduce the complexity and allow a better comparison.

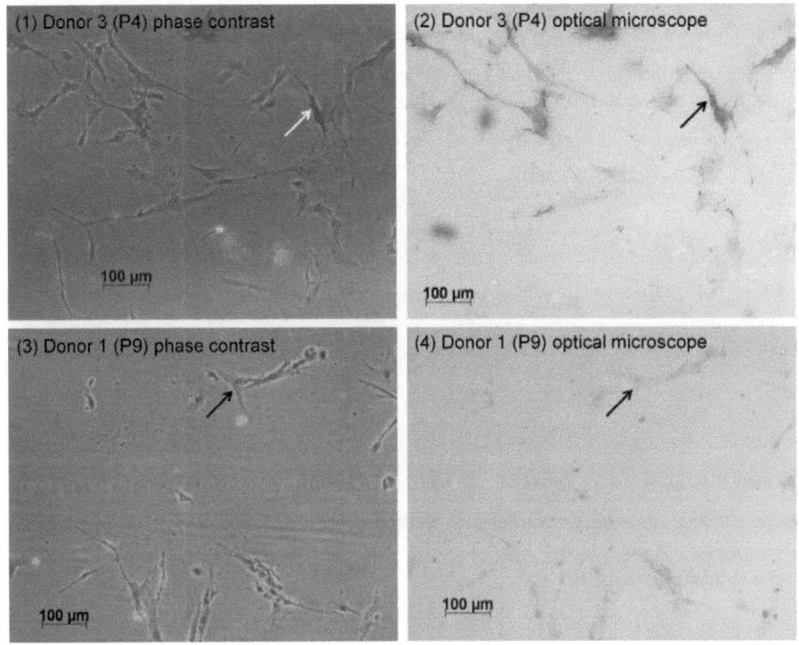

Figure 17: Cells from donor 3 (44, m) at passage 4 and donor 1 (18), m at passage 9 were stained for ALP. (1) and (3) Cells observed in phase contrast; (2) and (4) Stained cells observed in light optical microscope. The arrows indicate examples for ALP-positive and negative stained cells.

As osteoblast and fibroblasts originate both from multipotent mesenchymal stem cells and express the almost identical gene set, they are hard to differentiate.[41,154] Additionally primary human osteoblasts tend to dedifferentiate to fibroblasts after some passages even when cultured in osteoblast specific media. However, this process can be decelerate by osteogenic media additives, like ascorbic acid, and by the use of calcium free media, as fibroblasts need more calcium for growing.[148,155]

To distinguish between these two cell types we used the staining of ALP. In contrast to fibroblasts, bone cells are known to express ALP in high levels, because it plays a role in the matrix mineralization and for tissue-specific hormones, such as parathyroid hormone, as well as many other hormones, cytokines and growth factors.[148,156,157] Compared to the IHC (immunohistochemical) staining of osteocalcin or the performance of a polymerase chain reaction (PCR) for Cbfa1 the detection of ALP a straightforward procedure.

To obtain a negative control, we stained the HFIB fibroblast cell culture for ALP as illustrated in Figure 18. We cannot see any phosphatase staining for this cell line. This emphasizes the differences in ALP activity between osteoblasts and fibroblasts.

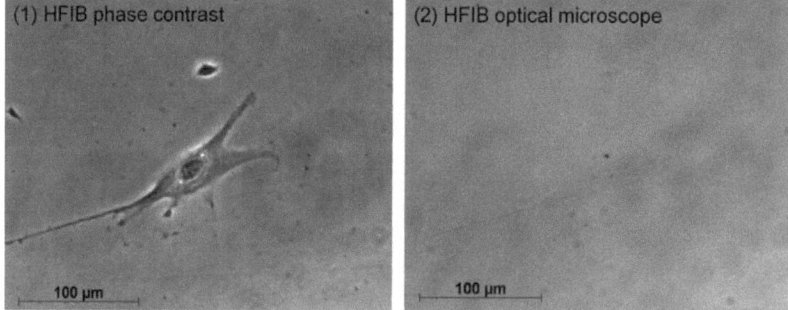

Figure 18: ALP staining on fibroblasts. HFIB fibroblast cell line was used to confirm that fibroblasts do not express ALP. (1) Cells observed in phase contrast (2) Stained cells observed in light optical microscope.

2.1.3 Confirmation of αvβ3 expression on osteoblast membrane

In the next step we proved the presence of integrin receptors expressed by our pooled osteoblasts. As mentioned before, the exact integrin pattern of osteoblasts remains unknown and during the "life" of an osteoblast the composition of integrin receptors on the membrane changes, however, the particular formation remains unclear.[73,78] However, we can state that bone cells express the following integrin subtypes: αv, α4, α5, and β1, β3, and β5.[43,80] Siebers et al. detected also α1-3 and α6 on the membrane of osteoblasts.[73] Reasons for these variations originate in the detection technique (antibody specificity), the fixation technique of the cells, the conditions for the immunocytochemistry, and the source of the cells.[78] As the peptides investigated in this study are highly selective for αvβ3, we concentrated on this type of integrin. In Figure 19 we demonstrated the positive staining of our pooled osteoblasts not only for αvβ3, but also for the subtypes αv and β3 and therefore verify that the cultured osteoblasts express αvβ3 on their cell membrane.

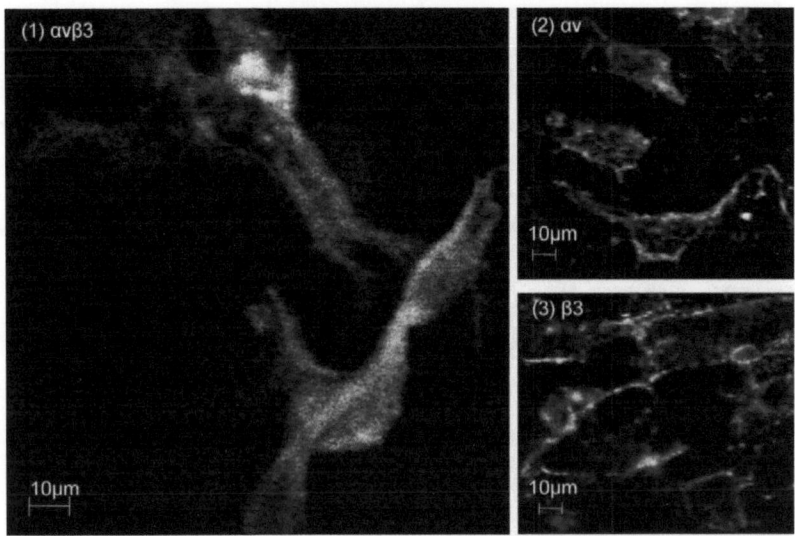

Figure 19: Immunofluorescent integrin staining for (1) αvβ3, (2) αv, and (3) β3.

We can state that we successfully obtained a cell pool containing primary human osteoblasts of six different donors. The cells of this pool express the common osteoblast characteristic markers and further αvβ3 receptors are present on the cell membrane. All cell experiments were performed with cells below passage 6 to avoid dedifferentiation of the osteoblasts to fibroblasts and challenge the reproducibility.

2.2 Investigated surfaces vary in terms of roughness

The objective of this section is to emphasize the importance of surface analysis and characterization in order to study osteoblast adhesion on these surfaces. The cellular system is very complex and cannot be completely standardized; therefore, the definition of pertinent parameters of the surface is even more important. It is crucial to be aware of the roughness parameters at all scales as well as the organization and structure of roughness.[97,158,159]

For our experiments we choose Ti6Al4V disks obtained from the company Biomet. The material quality of these disks is approved according to clinical standards for medical implants. To reduce the complexity only this titanium alloy and no other implant material was used. Different cell adherence has been reported on titanium and its alloys than for example on cobalt-chrome.[78,160] Anselme et al. stated recently that bone cells are more sensitive to surface topography than to material composition.[158]

We selected three surface textures with distinct surface treatments: trimmed, matt finished, and sandblasted. The roughness values are listed in Table 2 and were obtained with a profilometer as described in the materials and methods section. Sandblasted surfaces show an average roughness of 3.24 µm, significantly higher than trimmed surfaces (0.74 µm) or matt finished surfaces (0.66 µm).

Table 2: Roughness parameters for Ti6Al4V disks with three different surface textures.

Roughness parameter in µm	Trimmed	Matt finished	Sandblasted
S_a [a]	0.74	0.66	3.24
S_z	12.16	10.02	49.64
S_q	0.98	0.92	4.20
S_{sk}	0.34	1.09	0.12
S_{ku}	5.49	9.04	4.88
S_p	7.48	6.70	30.28
S_v	12.16	3.32	19.36

[a] S_a, average roughness; S_z, reflects peak height; S_q, root mean square; S_{sk}, asymmetry of the height distribution or surface skewness; S_{ku}, sharpness of the surface height distribution or surface kurtosis; S_p, largest peak height; S_v, largest valley depth.

The structure of different Ti6Al4V disks is shown in SEM images in Figure 20. On trimmed surface the circular grooving is evident in all magnifications, whereas the image of the matt finished disk looks very plane in the lowest extension, but the structure with irregular valleys and peaks gets visible at higher magnifications. The third texture, sandblasted surface, is without doubt the roughest one in this line. In all images we see the anisotropic and sharp-edged structure found after sandblasting Ti6Al4V surfaces with EK 54 corundum particles. Most efforts of adhesion improvements between bone tissue and Ti6Al4V base on surface roughening by the use of blasting.[22,161]

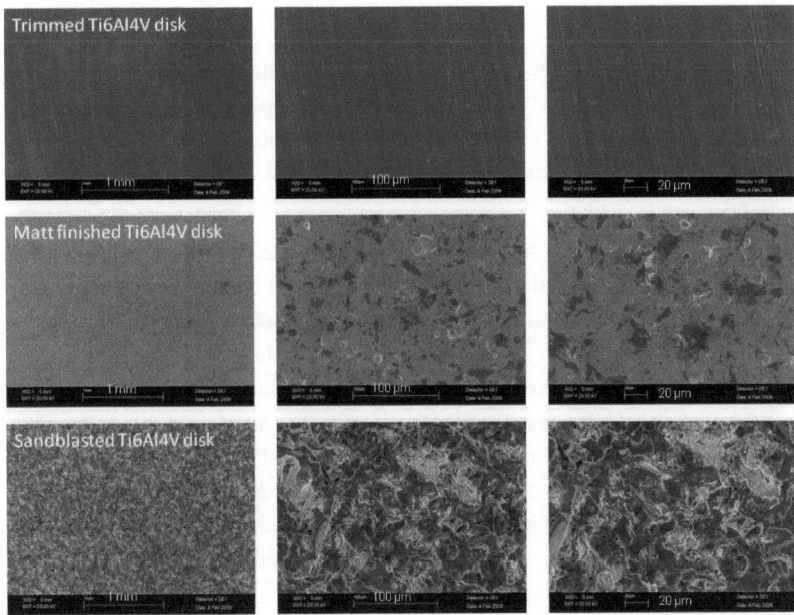

Figure 20: SEM images of trimmed, matt finished and sandblasted Ti6Al4V disks in different magnifications.

Trimmed and matt finished surfaces were chosen to investigate differences in the shape and orientation of the surface, despite the fact that the average roughness is rather similar. Roughness alone, however, is unable to distinguish peaks and valleys.[159]

Observations in a reflected-light microscope indicated that trimmed Ti6Al4V disks have an isotropic surface with circular grooving, whereas the matt finished structure is anisotropic. Both surfaces display for most roughness parameters similar values. Only the dimensions of S_v (maximum depth of valleys), S_{sk} (surface skewness), and S_{ku} (surface kurtosis) differ considerably. All three parameters display the distance between the lowest/ highest point and the assessment length average.[162] Trimmed surfaces have less, but higher amplitudes, whereas matt finished disks are overall more turbulent only in lower amplitude. The third kind of disks posses a sandblasted surface that was shown to be preferred by osteoblasts.[3,6,73] For this anisotropic surface, a 4- to 5-fold higher roughness was measured; hence the real surface area for these disks is higher compared to trimmed and matt finished disks. The exact surface area is hard to determine, as the precise radius is unknown due to the roughness.

In Figure 21 we display the profiles of a glass slide and the three different Ti6Al4V disks. The increasing amplitude demonstrates the enlarging roughness value. On trimmed and matt finished surface the distance between the highest and the lowest peaks is 19 µm and 10 µm, whereas on sandblasted amplitudes up to 50 µm are detected.

Different properties of the applied material are crucial for the interaction between cells and implant. The chemical composition needs to be nontoxic; at least bioinert, but better bioactive, meaning cells prefer this composition towards other materials. Additionally, it has been shown that the chemical composition influences the expression of membrane proteins, especially of integrins.[78] The wettability and thus the surface free energy describe another factor that influences the first contact of proteins and cells on the surface. This parameter constitutes a fundamental property of solid surfaces, but it is also influenced both by the chemical structure and the surface topography. When a surface is exposed to water or media the findings are ambivalent depending on the surface texture, both the hydrophilicity and the hydrophobicity of the surface can be increased.[163]

Surface roughness is the most important parameter affecting the adhesion process. Even after the cells are adhered, the roughness has impact on the proliferation and the differentiation of the cells, because the signaling within the cell are controlled by forces from the ECM.[73,164]

We can assert that the literature is consistent on the relevancy of roughness, but inconsistent on whether rough or smooth is the better surface for osseointegration. Several statements can be found both for the benefit of rough surfaces[165-167] and for smooth surfaces.[168-170] Anselme et al. found a better spreading on surfaces with low roughness amplitude, but a higher adhesion power on rough isotropic surfaces, concluding a higher sensitivity of osteoblasts towards the organization and morphology of the roughness than to its amplitude.[171] When the macro architecture of the applied material presents a morphological feature greater than the scale of osteoblasts, the cell might be unable to sense the roughness due to proportional reasons. Conversely, this means a surface with a roughness value in nanometer range can affect cell adhesion to a stronger extent than a surface with higher roughness parameters, because the roughness is in the dimension of the cell.[100,170,172]

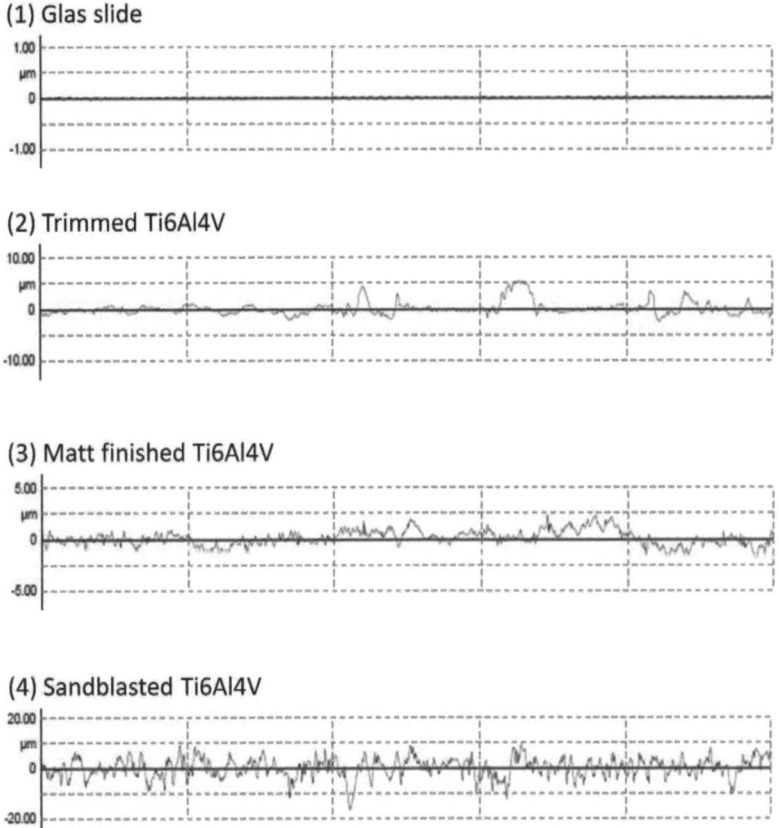

Figure 21: Profile of used materials. (1) Glass slide and Ti6Al4V with different surface properties: (2) Trimmed, (3) Matt finished, and (4) Sandblasted. Scan speed = 0.8 mm/sec; Length of measured line = 3.98 mm; N.B. different scale of ordinate.

Summing up, the major determined parameter for cell adhesion is the topography of the implant surface, and therefore its roughness. By choosing these surfaces with different roughness values, we intended to perform the tests on isotropic and anisotropic surfaces and additional on varying levels of roughness.

2.3 Binding assays with Integrins verify affinity of RGD peptides

In this chapter we present the results of the cell free assays. For these experiments an ELISA-like assay with isolated integrins was used and we demonstrated the specific binding of RGD peptides for the target integrin, namely $\alpha v\beta 3$. Additionally we investigated the integrin binding on different surface textures, namely trimmed, matt finished and sandblasted Ti6Al4V disks. Further, we adapted this assay for another integrin subtype, namely $\alpha 5\beta 1$.

All molecules in this work were synthesized within the group of Prof. Kessler under the supervision of Dr. Carlos Mas-Moruno (RGDfK by Mona Wolff, peptide **1 - 7** by Dr. Carlos Mas-Moruno, peptide **8 - 11** by Stefanie Neubauer and Dr. Carlos Mas-Moruno, compound **I** to **V** shown in section 2.3.5 by Florian Rechenmacher).

2.3.1 Optimization of integrin binding assay for $\alpha v\beta 3$

Before exploring the effect of peptide coating onto surfaces with various levels of roughness, we investigated the optimal chemical structure of the coating molecules and the best conditions for the assay. The chemical structures of the $\alpha v\beta 3$ specific peptides used in this study are given in Figure 22.
We chose the cyclic RGD peptide **1** as it efficiently enhanced osteoblast adhesion on Ti6Al4V in a previous work.[130] This molecule is based on the sequence c(-RGDfK-), which displays affinity for the integrins $\alpha v\beta 3$ and $\alpha v\beta 5$ in the low nanomolar range and selectivity against the platelet receptor $\alpha IIb\beta 3$.[43,173] As anchoring system this compound contains a branched tetraphosphonic acid that ensures a tight and stable binding of the peptide to titanium.[42,109] To compare the effect of this anchor, the tetraphosphonic acid was substituted by a thiol group in peptide **2**. To study the importance of the spacer, the tetraphosphonate-functionalized peptide was synthesized without the Ahx linker (peptides **3**). In addition, for the thiol-functionalized peptide the analogues without and with two, one aminohexanoic acid (Ahx) linker (peptides **4**, **5**, and **6** respectively) were prepared. Finally compound **7** was produced to measure the impact of a single phosphonic group on the coating activity.

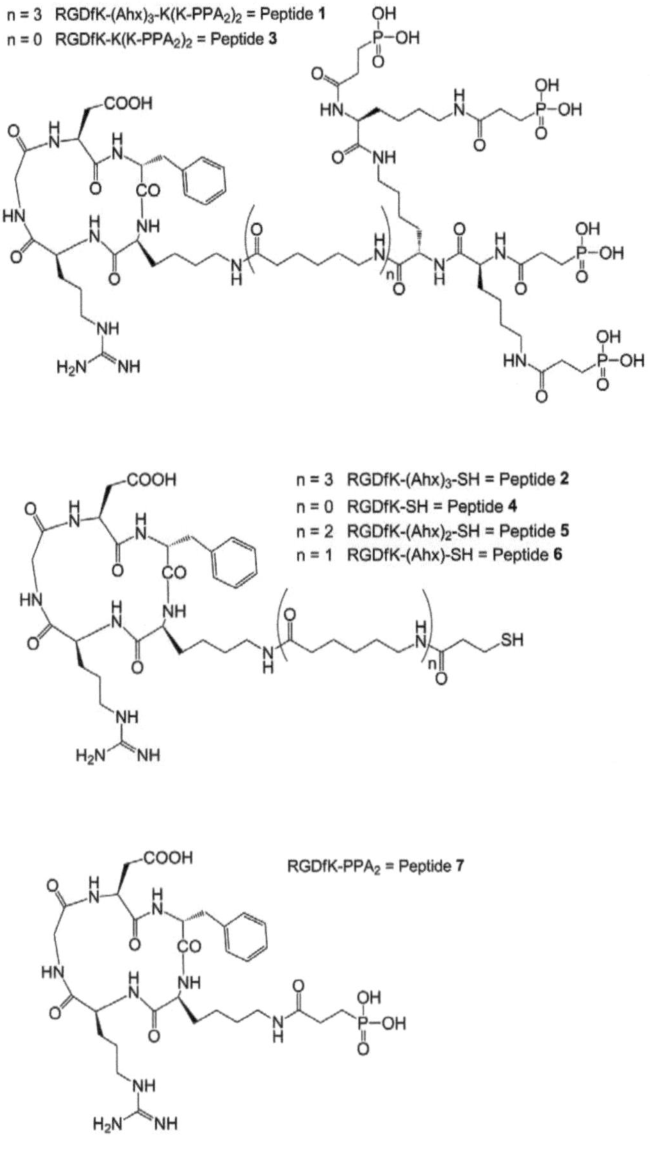

Figure 22: Chemical structures of the modified RGD cyclic peptides used for coating on Ti6Al4V disks. The peptides were synthesized and characterized under the supervision of Dr. Carles Mas-Moruno.

In the first step we analyzed the linear correlation between RGD peptide concentration and the absorption value of biotinylated αvβ3 at 492 nm. To test the effect of coating molecules we used a modified ELISA-like protocol (see the materials and methods section). The principle is sketched in Figure 23. After coating of the desired surfaces with RGD peptide, uncoated positions are blocked with BSA. The next step involves the incubation of biotinylated integrin, which binds selectively to the coated RGD peptides. The assay proceeds with the addition of avidin horseradish peroxidase (HRP) conjugate. This reaction takes advantage of the high affinity displayed by streptavidin to bind biotin. Finally, the oxidation of a suitable substrate by HRP, using H_2O_2 as oxidizing agent, yields a strong colorimetric signal that can be spectrophotometrically measured at 492 nm. Such a testing system is not directly quantifying the amount of coating material, but exhibits the amount of accessible RGD peptides for integrin binding. This information together with the exact amount determined by radio labeling methods (see Chapter 2.3.4) provides a better understanding of the accessibility of coated RGD peptide on the surface.

Figure 23: Principle of the integrin binding assay on surfaces coated with RGD peptides.

Initially we coated plastic surfaces with increasing concentrations of peptide **1** and measured the amount of αvβ3 binding (Figure 24). Up to 50 µM, the effect of peptide coating is negligible. At higher concentrations a dose dependent correlation between peptide concentration and integrin binding is observed. The higher the concentration of coating material, the higher is also the integrin binding until all binding sides are occupied and a specific threshold is reached.

Considering that the synthesis of peptides is laborious and costly, the lowest concentration possible should be coated, however, it is mandatory to ensure enough peptides for the binding of all possible sites on the surface. We can state that a concentration of 100 µM peptide in the coating solution is suitable for our cell-free assays. Hence, the experiments in the following were performed with this concentration. This value is comparable to concentrations used by others. Auernheimer et al. investigated the effect of peptide concentration in terms of peptide attachment onto the surface and on integrin binding.[130] They applied a peptide concentration of 100 µM in the coating solution for most of their assays.

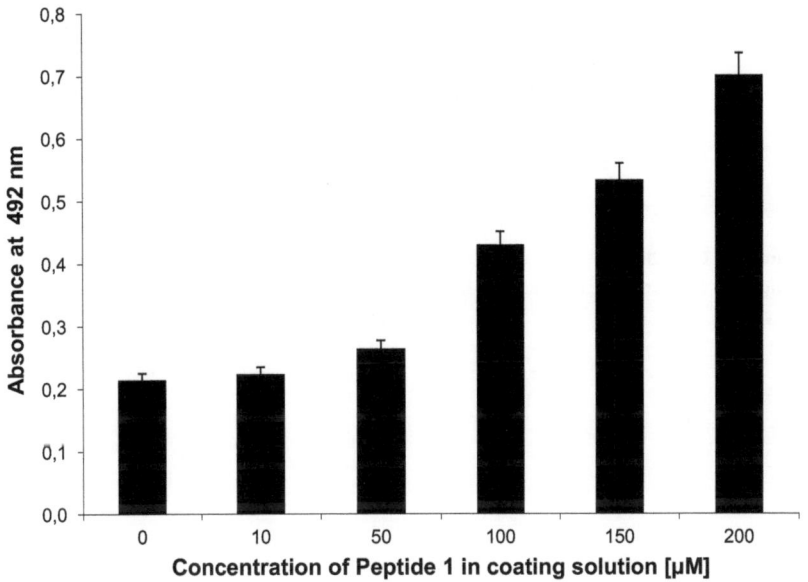

Figure 24: Detection of αvβ3 integrin on plastic coated with different concentrations of peptide 1 in the coating solution (100 µL). The amount of attached integrin was quantified by the established ELISA-like assay.

In Figure 25 we show the results of a similar experiment performed with different concentrations of Cilengitide as coating molecule.[109] The αvβ3 antagonist Cilengitide serves as negative control in order to prove the principle of this assay. The chemical structure of Cilengitide (Figure 26) displays a cyclic RGD peptide, based on the peptide cyclo(-RGDfV-), without any spacer or anchor motif. This molecule is unable to bind to any surface.

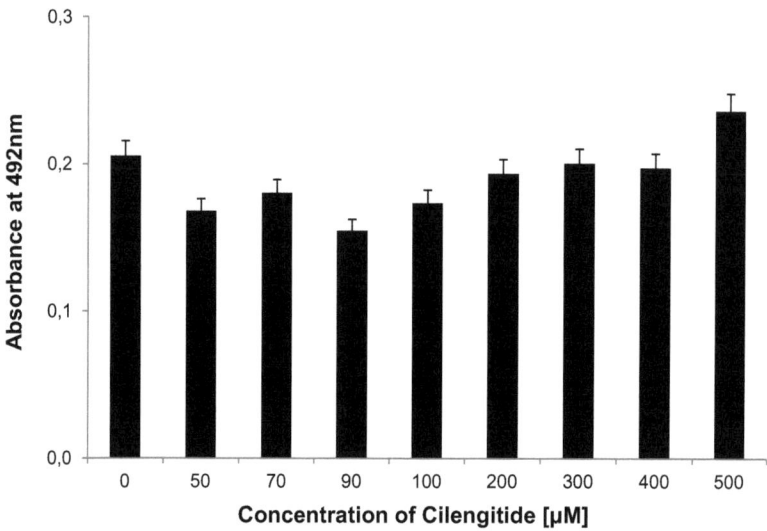

Figure 25: Detection of αvβ3 integrin on plastic coated with different concentrations of Cilengitide. The amount of attached integrin was quantified by the established ELISA-like assay.

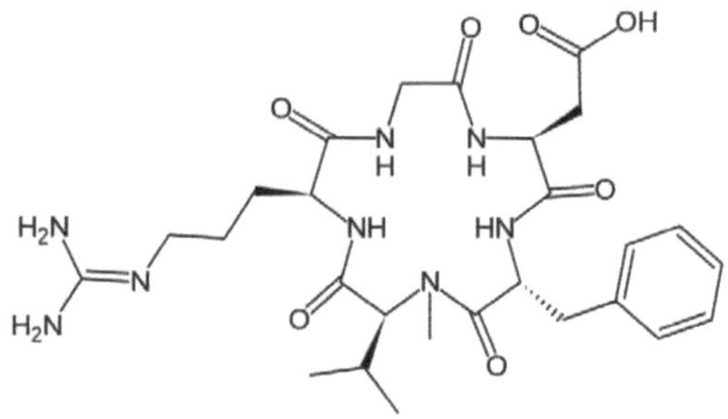

Figure 26: Chemical structure of Cilengitide (cyclo-[RGDfN(Me)V])

This αv-selective peptide was synthesized in 1995 within the group of Prof. Kessler in collaboration with Merck (Darmstadt) in order to combat aggressive cancers, like glioblastoma. By blocking the integrin receptors of cancer cells, angiogenesis can be inhibited and therefore the development of metastases is suppressed.[174] The drug is currently in clinical phase III for glioblastoma and the European Medicines Agency has granted Cilengitide the orphan drug status.[109,175]

In this experiment (Figure 25) no increase in absorbance is detected independently of the concentration of Cilengitide. Thus the head group of the peptide without any anchor and spacer is not able to present the RGD motif in a way that integrin binding can be stimulated. In that case also cell adhesion is unlikely to be supported.

In a next step we further optimized the integrin concentration for this assay. The modified ELISA-like assay was performed on the plastic surface of a multiwell plate coated with peptide **2** in different concentrations (1 and 100 µM) and two concentrations of biotinylated αvβ3 (5 and 10 µg/ ml) were used. As shown in Figure 27 we obtained a slightly stronger signal in the experiments with the higher integrin concentration. Although the difference is not strong, we decided to continue the further experiments with an integrin concentration of 10 µg/ ml.

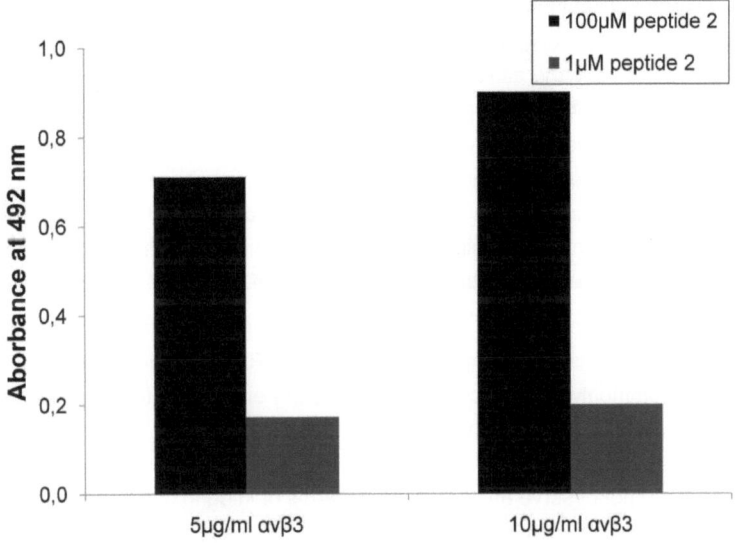

Figure 27: Optimization of αvβ3 concentration for the modified ELISA test system. Peptide 2 dissolved in PBS was coated overnight at different concentrations (1 and 100 µM). Two concentrations of biotinylated αvβ3 were used: 5 and 10 µg/ml.

2.3.2 αvβ3 binding on RGD coated trimmed Ti6Al4V disks

In the following experiments, we coated different peptides on trimmed Ti6Al4V disks in order to investigate the chemical structure of these peptides as coating molecules. All peptides applied consist of a selective ligand, a spacer unit and an anchoring system; in this study the selective ligand motif "RGDfK" is maintained for all peptides and we concentrated on variations of the anchor and spacer units and their impact on integrin binding.

The protocols for coating the disks and the performance of the assay are described in the materials and methods section. In Figure 28 we show that coating of RGD peptides featuring a distinct spacer and/or anchor unit (peptide **1**, **2**, and **3**) significantly enhanced αvβ3 binding compared to control disks (PBS-treated or uncoated). An overview of the differences within the peptide structures applied in this test is given in Table 3. Both anchor systems, tetraphosphonate (peptide **1**) and thiol (peptide **2**), showed almost the same potency for integrin binding. For the peptide containing a tetraphosphonate anchor (peptide **3**) the Ahx spacer was not required for gaining integrin binding. However, it is important to have a tetra-phosphonate anchor to retain the activity since compound **7** showed little effects in promoting αvβ3 binding. On the contrary, although one thiol group and the Ahx spacer is enough to keep the activity, without spacer the peptide with thiol anchor looses all activity (peptide **4**). The binding of αvβ3 to peptide **1** was inhibited in the presence of Cilengitide.[109] This documented the specificity of the enhanced integrin binding due to the integrin-RGD-interaction. Also, the use of unmodified cyclic peptide **c(RGDfK)** showed little effect in integrin binding confirming that the presence of an anchor unit is crucial for the correct presentation of the RGD motif.

Anchoring groups are intended to provide a strong and stable binding of the peptides to the implant material. In this regard, the binding of RGD peptides to titanium and its alloys has commonly been accomplished by using either thiol functionalities[31,94] or phosphonic acids.[80,123,176] Although it has been described a higher efficiency of phosphonates compared to thiols in the binding of peptides to titanium,[133,176] in our studies we did not observe any difference in terms of integrin binding affinity between these two anchors (Figure 28, peptides **1** and **2**).

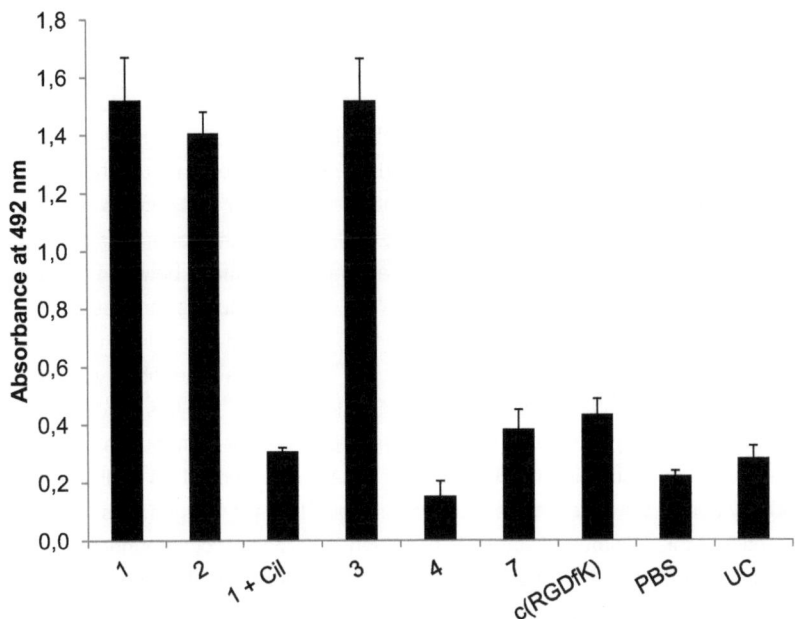

Figure 28: Binding of αvβ3 integrin to trimmed Ti6Al4V disks coated with RGD peptides (1 to 4, 7 and c(RGDfK)) was measured using the modified ELISA. Peptides were dissolved in PBS and were coated overnight at a concentration of 100 µM. The binding of αvβ3 to peptide 1 was inhibited in the presence of the super potent integrin antagonist Cilengitide (Cil, 27 µM). PBS solutions without peptide and an uncoated disk (UC) were used as control.

Table 3: Overview of the peptide structures applied in the test illustrated in Figure 28.

Peptide	Anchor	Spacer	Additive
1	$K(K-PPA_2)_2$	$(Ahx)_3$	-
2	SH	$(Ahx)_3$	-
1 + Cil	$K(K-PPA_2)_2$	$(Ahx)_3$	Cilengitide
3	$K(K-PPA_2)_2$	-	-
4	SH	-	-
7	PPA_2	-	-

The effect of the spacer units was also investigated. The influence of the Ahx$_3$ unit on integrin binding was insignificant for the tetraphosphonate peptide (**1** and **3**, Figure 28); however, its effect was dramatic for the biological activity of thiol peptide (**2** and **4**, Figure 28). A closer view of the chemical structure of each compound revealed important differences that can explain this discrepancy (see Figure 22). The presence of two branching units of lysine (Lys), required to construct the tetravalent phosphonate anchor, locates the RGD motif at a longer distance to the Ti6Al4V surface compared to the single monovalent thiol group. In this regard, several studies have already described the importance of a minimum distance between surface and ligand on integrin-mediated cell adhesion.[57,94,110,122,129] Studies for a series of acryl peptides coated on poly(methyl methacrylate) described a minimum distance of 3.5 nm for an optimal osteoblast adhesion.[110] This was also observed for thiol-functionalized peptidomimetics binding to Ti6Al4V disks.[94] In this study a reduced activity in mediating osteoblast adhesion was observed for compounds bearing only a mercaptopropionic acid unit compared to an Ahx$_3$-Cys unit. This finding was most probably due to the lower accessibility of the ligand for the integrin. It is obvious that in peptide **4** this minimum distance is not given, whereas for compound **3**, the double Lys unit behaves as a "pseudo-spacer" which provides the required distance between the RGD motif and the Ti6Al4V surface. The importance of this Lys branching unit was also evident for peptide **7**, which displayed no significant activity.

Therefore, for $\alpha v\beta 3$-binding studies on Ti6AlV4 surfaces, both thiol and phosphonic acids can be used as anchors as long as a minimum distance between the RGD binding sequence and the anchoring moiety is provided.[177]

2.3.3 $\alpha v\beta 3$ binding on RGD coated Ti6Al4V disks with different surface roughness

To study the effect of peptide coating on different topographies, three surface textures with distinct roughness values were selected (see Chapter 2.2). These Ti6Al4V surfaces were coated with peptides **1** and **2**, respectively, and the binding of $\alpha v\beta 3$ was measured as described in the materials and methods section.

Coating with peptide **1** that has a tetraphosphonate as anchoring system showed an enhancement in $\alpha v\beta 3$ binding for all surfaces when compared to control disks, the result of the experiment is illustrated in Figure 29. The highest enhancement was observed for the smooth trimmed surface (higher than 10-fold), compared to matt finished (6-fold) or sandblasted surface (7-fold).

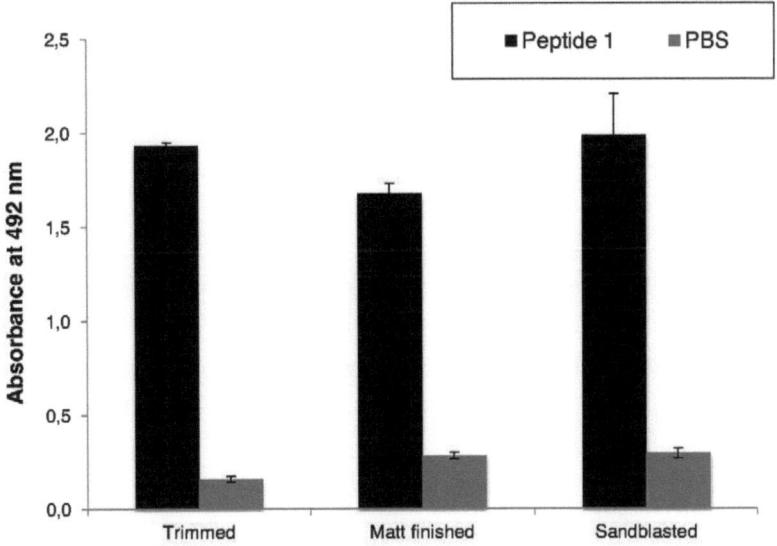

Figure 29: Binding of αvβ3 integrin to Peptide 1 coated Ti6Al4V disks with different roughness properties (trimmed, matt finished and sandblasted) was determined using a modified ELISA. Ti6Al4V disks were coated overnight with 100 μM of peptide 1 (RGDfK-(Ahx)$_3$-K(K-PPA$_2$)$_2$, black bars) or treated with PBS (grey bars).

A similar trend was observed for disks coated with peptide **2** possessing a thiol anchor; however, the presence of this peptide boosted the integrin binding to a lower degree. The results are shown in Figure 30. A 6-fold increase was observed on trimmed, 3-fold on matt finished and 4-fold on sandblasted surface, respectively.

When comparing these results with each other, we can state that for both peptides the highest enhancement of integrin binding was achieved on trimmed surface. On matt finished and on sandblasted disks both, the peptide with the tetraphosphonate and the one with the thiol anchor, promote a similar integrin binding, although the level of enhancement for peptide **2** is lower.

We postulate a better accessibility of coated RGD peptides on trimmed surface compared to matt finished or sandblasted textured disks, a more detailed discussion of this topic can be found in the next section (Figure 32).

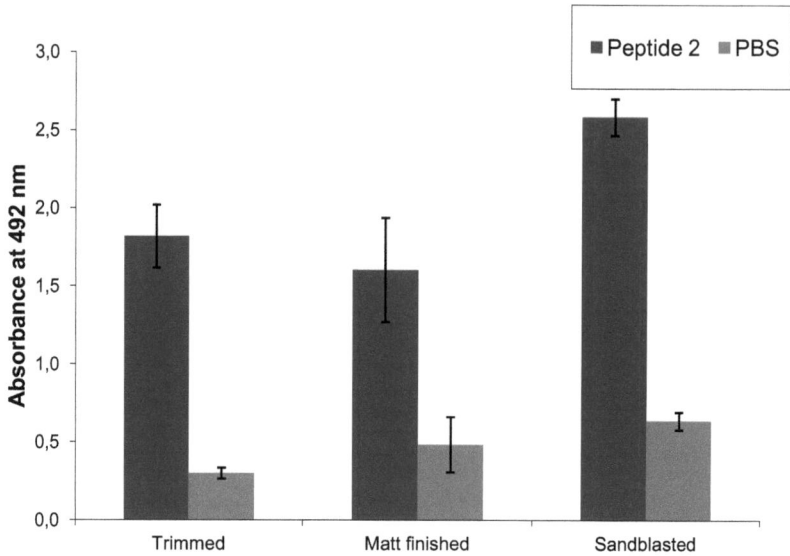

Figure 30: Binding of αvβ3 integrin to Peptide 2 coated Ti6Al4V disks with different roughness properties (trimmed, matt finished and sandblasted) was determined using a modified ELISA. Ti6Al4V disks were coated overnight with 100 µM of peptide 2 (RGDfK-(Ahx)$_3$-SH, dark grey bars) or treated with PBS (grey bars).

2.3.4 Quantitative analysis of RGD peptides bound on Ti6Al4V disks by radio labeling

The experiments in section 2.3.3 demonstrated that surface roughness influences peptide accessibility and thus integrin binding. To further explore the influence of surface roughness, we used radio labeling studies to determine the amount of RGD peptide **1** and **2** bound to different surface structures.

For this purpose, analogues containing a D-Tyr instead of a D-Phe were synthesized (peptides **8** and **9**) and labeled with ^{125}I. The chemical structures are given in Figure 31. These peptides were coated onto Ti6Al4V disks following the coating conditions as described in the materials and methods section. Radioactivity on the surface was determined to analyze the number of peptides bound to each surface.[173]

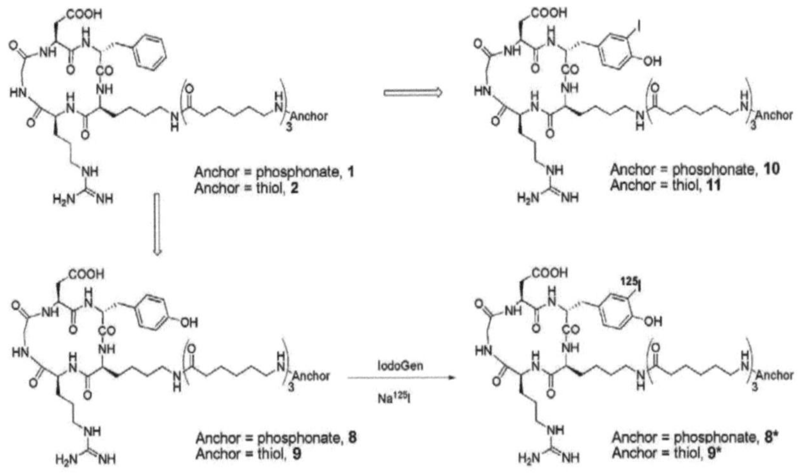

Figure 31: Design of RGD modified peptides for radio labeling studies. The D-Tyr-containing analogues 8 and 9 were radio labeled with ^{125}I to perform radioactive measurements. The cold peptides 10 and 11 were used as controls.

In Table 4 we listed the data obtained for the coating of ^{125}I-tetraphosphonate peptide 8*. We observed a rather similar amount of bound peptide on both trimmed and matt finished disks (0.4 pmol/ cm^2 and 4.5 pmol/ cm^2, respectively). In contrast, the amount of peptide bound to sandblasted disks was prominently higher (27 pmol/ cm^2).

Table 4: Average roughness and amount of ^{125}I-labeled peptide 8* bound per disk.

Surface	Average roughness (S_a) (µm)	Peptide bound per disk (pmol/cm^2)[a]
Trimmed	0.74	6.4 ± 2.2
Matt finished	0.66	4.5 ± 0.5
Sandblasted	3.24	27 ± 12

[a] Surface area of disk equal to 0.785 cm^2 (calculated with a diameter of 1 cm for each disk, independently of the roughness)

One reason for the higher amount of peptide found on sandblasted surfaces bases on the fact that because of the roughness the actual surface area of these disks is higher than 0.785 cm^2, though the precise surface area is hard to determine. However, the results of the radio labeling studies are in contrary to our previous findings (see chapter 2.3.3), in which peptide **1** coating on sandblasted surfaces did not result in a 4-fold increase of integrin binding on trimmed surface or a 6-fold increase on matt finished, respectively. As shown in section 2.3.1 we observed a linear correlation between the peptide concentration and the bound amount of integrin, thus we would expect a higher integrin binding on rougher surfaces compared to smooth ones. Similar observations were also made for the thiol-functionalized peptide, as peptide coating enhanced the integrin binding only by a factor of 1.4 on sandblasted versus trimmed surfaces.

Therefore, we assume that peptides on trimmed (smooth) surface topography were more accessible for the integrins than on sandblasted (rough) disks. We hypothesize *that on a rough surface some molecules are hidden in the "valleys" of the disk and* thus may not be reached by integrins, whereas on a trimmed surface most peptides are accessible to the integrins. These assumptions are schematically represented in Figure 32.[177]

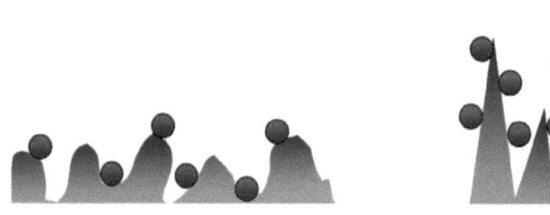

trimmed = smooth   sandblasted = rough

Figure 32: Schematic representation of the accessibility of RGD peptides on titanium disks with distinct roughness properties. Green spheres represent RGD peptides accessible to the integrins, whereas red spheres indicate positions hindered and not accessible.

The density studies of Healy and Rezania on quartz surfaces give another argument. At high-ligand densities a plateau is reached and in their experiment this concentration was around 0.67 pmol/ cm^2,[136] as for higher concentrations (e.g. 3.8 pmol/ cm^2) no further increase was detected.[137] They assume at 0.67 pmol/ cm^2 a ligand density of ≥ 3,600 RGD molecules/ μm^2 (assuming 2,000 integrin receptors per μm^2).

When doing the calculations, based on the amount of coated peptide determined by our radio labeling studies, the total number of RGD molecules per μm^2 is ~ 40,000 for trimmed, ~ 30,000 for matt finished and ~ 160,000 for sandblasted (calculated by multiplying the peptide concentrations per cm^2 from Table 4 with the Avogadro constant). Using the assumption from Healy and Rezania that each μm^2 possess 2,000 integrin receptors, the ratio of RGD peptide per integrin is 20 for trimmed, 15 for matt finished and 80 for sandblasted, respectively. Although we cannot determine the saturation threshold for our set-up, we hypothesize that the saturation level is certainly reached when the coated sandblasted surface presents 80-times more ligands than receptors available on the cell surface. Therefore the increase of integrin binding cannot be expected to be linear for the three different surfaces.

2.3.5 Development of the assay for α5β1

In medical chemistry it is a major challenge to gain more insights about the specific role of integrin subtypes. There is a wide range for possible applications of selective peptides either to induce specific cell adhesion, like on implants in orthopedics/dentistry and on stents in cardio surgery, or to target specific cancer types for diagnostics and therapy. One of the first steps in developing potential new peptides is to test the compounds with the established system for their activity and selectivity towards the distinct integrins.

In the group of Prof. Kessler distinct thiol- and phosphonate-functionalized α5β1-selective ligands are synthesized to investigate different research areas. In the first step we adapted the established assay for α5β1 binding. The protocol was modified as described in the materials and methods section.

The structures of the tested compounds are given in Figure 33. The results obtained with the adapted assay for α5β1 binding are shown in Figure 34. The absorbance, and therefore the integrin binding, was increased significantly by 2 of the 5 tested compounds: compound I by a factor of three ($p \leq 0.001$) and compound IV by a multiple of five ($p \leq 0.001$) compared to the signal maintained with PBS treated disks. The other three compounds did not raise the binding of α5β1 compared to the controls. The comparison of the anchor structures can explain these findings. One thiol group or a bivalent phosphonate group provides a sufficient anchoring system for the coating on Ti6Al4V.

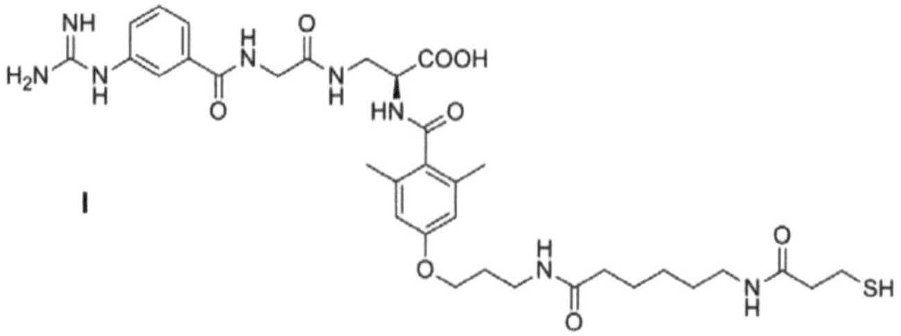

I

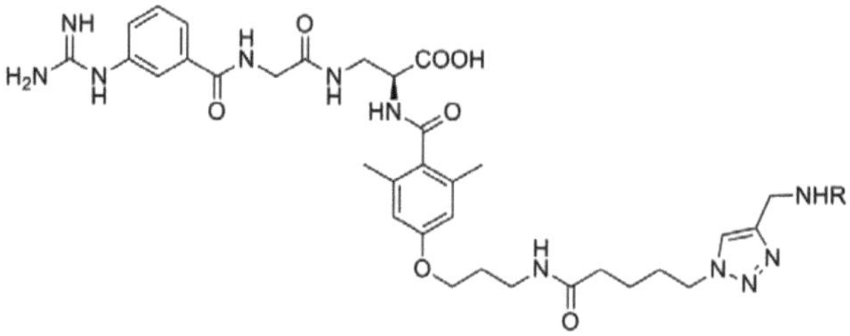

R=

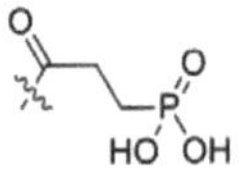

II

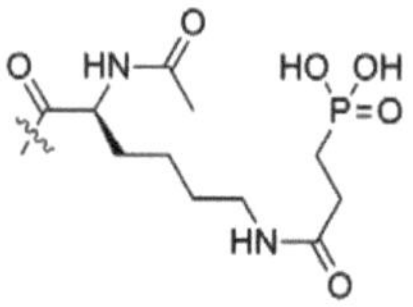

III

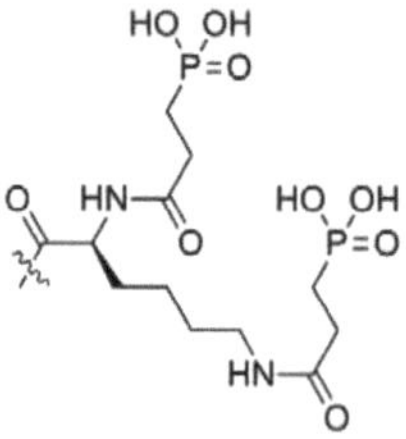

IV

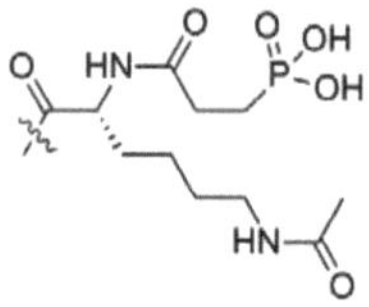

V

Figure 33: Chemical structure of α5β1-selective compounds I to V. The peptides were synthesized and characterized by Florian Rechenmacher.

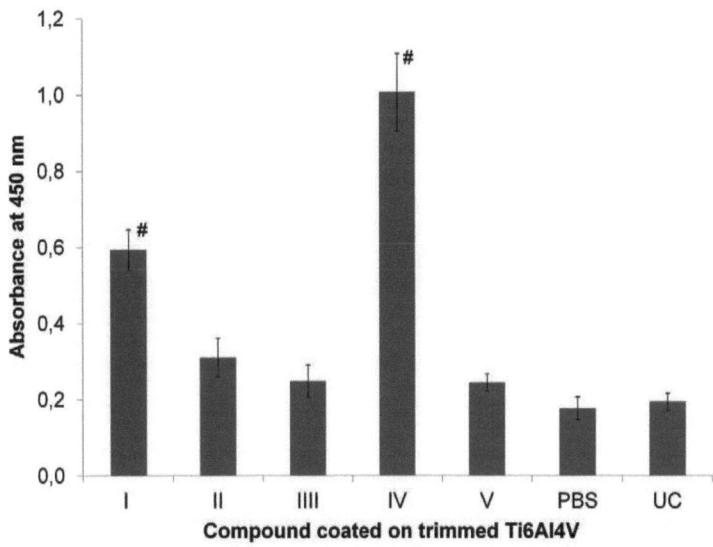

Figure 34: Binding of α5β1 to Ti6Al4V disks (trimmed) was determined using a modified ELISA. Ti6Al4V disks were coated overnight with 100 µM of different peptidomimetics (I- V), treated with PBS or uncoated (UC). Values are given as means ± standard deviations. Statistical differences with p ≤ 0.001 are indicated with the symbol (#).

The assay can also be used to compare and investigate the selectivity for each specific integrin subtype. In general the effectiveness of a compound or the antagonist drug potency is given by the half maximal inhibitory concentration (IC_{50}).[178] This value indicates what concentration of the substance is necessary to reduce a biological process, in this case integrin binding, by half.[178] E.g. the determined IC_{50} values for compound I are 9600 nM for αvβ3 and 0.86 nM for α5β1.

The result of an experiment examining the activity and selectivity is given in Figure 35. Compound I was analyzed in terms of α5β1 and αvβ3 binding. On the coated Ti6Al4V disk we detected a significant increase in the absorbance for α5β1, but no increase for the signal of αvβ3. This finding confirms the determined selectivity profile and demonstrates that the peptide is highly active and selective for α5β1, but not active for αvβ3 binding.

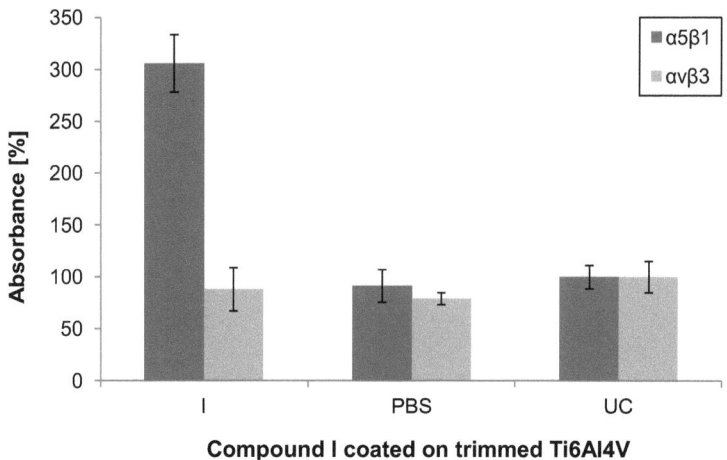

Figure 35: Binding of α5β1 and αvβ3 to Ti6Al4V disks (trimmed) was determined using a modified ELISA. Ti6Al4V disks were coated overnight with 100 µM of compound I, treated with PBS or uncoated (UC). The absorption detected at 450 nm for α5β1 and 492 nm for αvβ3 was normalized to the value measured on uncoated probes, so a comparison between the different assays is possible.

With the above described experiments, we stressed the potential of the developed ELISA-like assay for a reliable estimation of compound bound to the given surface. In order to test the influence of the anchor/ spacer system in detail a molecular toolkit was established, and particular peptidomimetics were synthesized systematically with different anchors and spacer units. Additionally, the established test system can be applied to analyze the peptide-surface-interaction on other surfaces, like e.g. stent material. So it is possible to investigate not only the coating material, but also the used surfaces.

2.4 Impact of RGD peptides and surface structure on the osteoblast adhesion process

In Chapter 2.1 to 2.3 we provided the base for experiments with cells. We assembled a cell pool, characterized the surfaces, and ensured an active and selective peptide-surface interface for integrin binding.

In Chapter 2.4 we focus on the investigation of the adhesion process on these surfaces. We study the influence of peptide and surface combinations in reference to enhance osteoblast adhesion. Other sections of this chapter examine the dependency of the obtained results in respect of the testing conditions. We stress the influence and importance of the BSA blocking step and the FCS addition in the assay media. In addition we analyze the time dependency of the adhesion process and detect the highest impact of peptide coating during the early adhesion phase. These effects are investigated with different methods analyzing the adhesion. Each approach concentrates on particular parameters and aspects of the adhesion process.

2.4.1 RGD peptide concentration for cell assays was optimized

In the first step we investigated osteoblast adhesion as a function of peptide **1** concentration in order to find the optimal coating concentration for the experiments performed with osteoblasts. We used the hexosaminidase assay to perform these experiments. Trimmed Ti6Al4V discs were coated with peptide concentrations ranging from 0.5 μM to 100 μM. The details of the test are described in the materials and methods section.

In Figure 36 we show the results of this experiment. Compared to the control disks, the amount of cells on disks coated with 0.5 μM of peptide **1** is more than doubled, when coated with 5 μM a 4-fold increase of the cell number is observed. A further enhancement of peptide concentration still raises the quantity of osteoblast although less intense. These findings are in agreement with the cell adhesion results of Auernheimer et al. and stress the benefit of coated RGD peptides for cell adhesion.[130]

As we performed all cell-free assays with a peptide concentration of 100 μM, we kept this concentration for our cell assays.

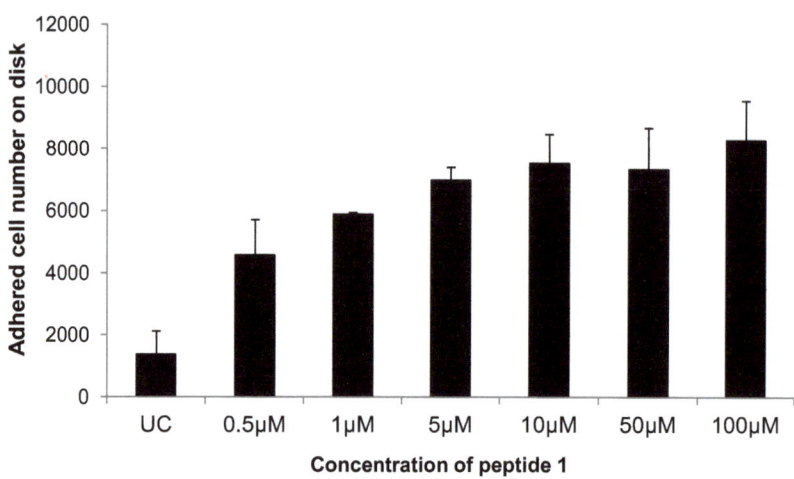

Figure 36: Analysis of different peptide concentration for the hexosaminidase adhesion assay. Peptide 1 dissolved in PBS was coated overnight at different concentrations (0.5- 100 µM). With the osteoblast pool at passage 5, the hexosaminidase assay was performed after 3 h adhesion time. The number of cells was calculated by the use of a standard curve.

2.4.2 RGD peptides coated on Ti6Al4V disks are non-toxic for osteoblasts

The cytotoxicity is a widely used parameter in pharmaceutical industry either to screen for cytotoxic compounds, e.g. for developing a therapeutic targeting cancer cells, or to test potential drugs for unwanted cytotoxic effects. One common used molecule for evaluating cell cytotoxicity is the lactate dehydrogenase test (LDH).[179]

To prove the non toxicity of our RGD peptides and the used surfaces, we tested the peptide coated on trimmed Ti6Al4V disks and the disks alone with an assay that detects LDH. The less LDH is detected, the more cells are vital on the surface. In Figure 37 we show that neither the uncoated nor the coated surfaces show any cytotoxicity on osteoblasts. For the positive control cells were poisoned with triton-X to ensure the accuracy of the test. Cells cultured in culturing plastic served as negative control.

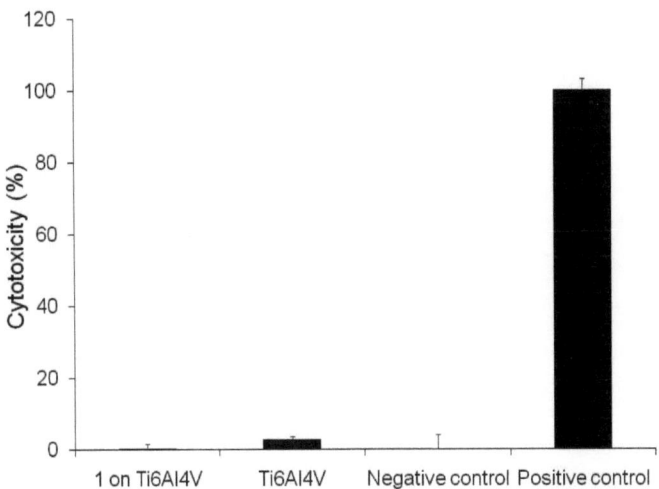

Figure 37: Peptide 1 was dissolved in PBS and was coated on trimmed Ti6Al4V disks at a concentration 100 μM overnight. The assay was performed with the osteoblast pool at passage 4; cells were treated with Triton-X (2 % in assay medium) for the positive control, as negative control cells cultured on culturing flasks were used.

2.4.3 Cilengitide inhibits binding of osteoblasts to Ti6Al4V disks

Cilengitide reduces the adhesion of cells presenting these integrin on their cell membrane, e.g. osteoblasts, in a concentration-dependent manner.[109] By blocking the αvβ3 receptors on bone cell membranes with Cilengitide, we show the dependency of osteoblast adhesion on this integrin.

The hexosaminidase test was performed with cells from two donors (D-1 and D-3), in media containing different concentrations of Cilengitide (0.1 μM and 100 μM). Additionally, the application time of Cilengitide was varied: first we added Cilengitide immediately and, secondly, cells were incubated for 4 h before the addition of the drug.

The results with immediate presence of Cilengitide in the cell suspension are shown in Figure 38 (1). The low drug concentration of 0.1 μM had no influence on the osteoblast behavior; however, 100 μM of Cilengitide lead to a complete repression of the adhesion process for cells of both donors.

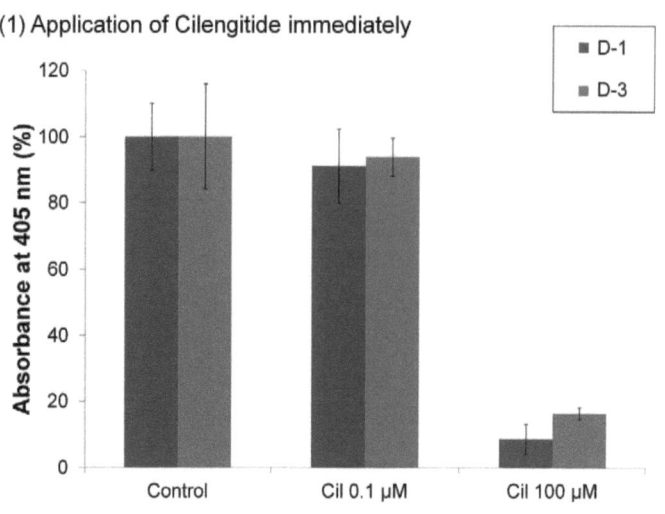

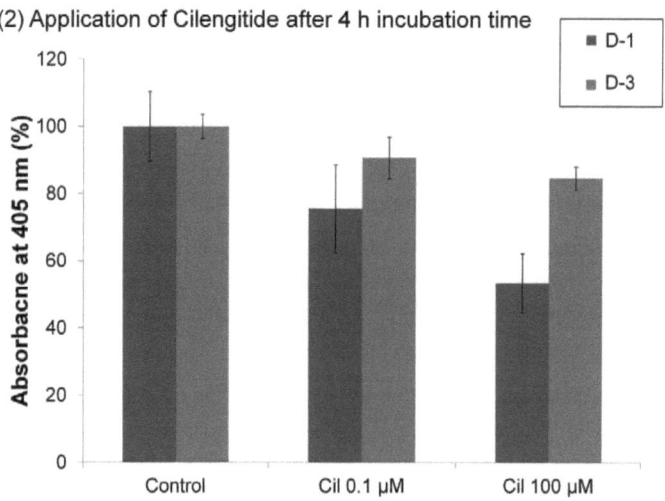

Figure 38: The hexosaminidase test was performed with osteoblasts from two donors (D-1 and D-3 at passage 4) on culturing plastic: (1) Cilengitide (0.1 µM and 100 µM) was immediately applied; (2) Cilengitide (0.1 µM and 100 µM) was applied after 4 h of incubation time. Extent of adhesion is given in per cent with control being set to 100 %.

Figure 38 (2) illustrates the results of the same set-up varying only in the application time of Cilengitide. Osteoblasts were first incubated for 4 hours before Cilengitide was added. Again 0.1 µM Cilengitide did not affect the cell adhesion significantly, but high concentration of the drug (100 µM) directed to a reduction of adhered cells, however, to a much lower degree.

The repression of cell attachment efficiency in the presence of Cilengitide demonstrates the strong dependency of osteoblast adhesion on $\alpha v\beta 3$. Once cells are adhered, an integrin antagonist, like Cilengitide, can hardly harm the process of adhesion. There are only free binding sides on the upper part of the cells and also during cell division integrins are disengaged, because for the cleavage cells need to detach from the surface. Both figures illustrate how individual primary cells react on the same treatment in a cell culture system. To minimize this aspect a cell pool of six donors was used for the further experiments.

In this experiment we used Cilengitide to show the dependency of the cellular adhesion process on RGD peptides and integrins. In cancer therapy the drug benefits from the highly upregulation of specific integrins, like $\alpha v\beta 3$ or $\alpha 5\beta 1$, on the endothelium during tumor angiogenesis.[109,180] By inhibiting the interaction between integrins and their ECM ligands, Cilengitide induces apoptosis in these cancer cells and suppresses the formation of new blood vessels. This effect is attenuated when the tumor is also irradiated, because in addition to the induction of cell death this treatment also causes an up regulation of $\alpha v\beta 3$ and Cilengitide may normalize the tumor vasculature.[109,181]

2.4.4 Cells prefer rough surfaces coated with RGD peptides for adhesion

In this section we demonstrated the influence of RGD coated disks with different surface textures on the adhesion process of osteoblasts.

Analyzing the amount of hexosaminidase as a measurement of adherent cell number explored osteoblast attachment efficiency. Ti6Al4V disks with distinct surface treatments (trimmed, matt finished and sandblasted), were coated with peptides **1** or **2**, and we analyzed the adhered cells after 3 h incubation time.

The results are illustrated in Figure 39. The exposure of peptide **1** to sandblasted surfaces resulted in the strongest cell attachment efficiency (82 % of the applied cells adhered). However, when compared to non-coated surfaces, most enhancement of osteoblast adhesion was found on trimmed disks: 2.4-fold increase for trimmed versus 1.7-fold enhancement for sandblasted. Peptide **2** had on all surfaces a weaker effect on osteoblast adhesion. On matt finished surfaces, the influence of peptide 1 on the cell attachment efficiency was much lower and almost negligible for peptide 2 compared to the uncoated surface.

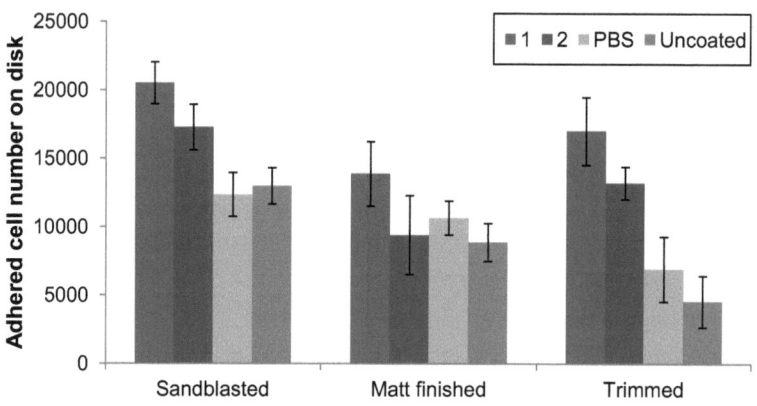

Figure 39: The osteoblast pool at passage 2 was passed to Ti6Al4V disks of different surface topography (trimmed, matt finished and sandblasted) coated with 100 µM of either peptide 1 or 2 as described. The total number of cells applied was 25,000 and after 3 h of incubation the number of adherent cells was estimated by hexosaminidase test.

These results indicate that the coating of surfaces with peptides has a relatively higher influence on trimmed surface topographies in comparison to sandblasted, which is in agreement with the values of $\alpha v \beta 3$ binding previously described in section 2.3.3.

We can also state that cells adhered most on sandblasted Ti6Al4V disks coated with peptide 1. The amount of osteoblasts was in general higher on this surface compared to the other topographies, such as trimmed and matt finished. For example, 50 % of all applied cells adhered on sandblasted disks without peptide coating. This presents already strong cell attachment efficiency, especially in comparison to trimmed surface where less than 16 % adhered on uncoated disks. This behavior had also been described in previous studies for osteoblast[12,182] and osteosarcoma cell adhesion[183] on titanium. An explanation for this observation is that on rough topographies the number of attachment sites for cells is higher compared to smooth surfaces. These adhesion sites are crucial, as they allow the formation of stable actin filaments and subsequent focal adhesions.

The influence of peptide coating in terms of osteoblast adhesion was higher on trimmed disks. The rather low adhesion value measured on control disks without peptide treatment (28 % of applied cells adhered on disks treated with PBS) was improved by a factor of 2.4 when these surfaces were coated with peptide 1. In comparison to that, a less pronounced enhancement on cell adhesion (only by a factor of 1.7) was observed on sandblasted disks. We have shown in our cell-free assays that the effect of peptide coating in promoting $\alpha v \beta 3$-binding on rough Ti6Al4V surface is less significant compared to smooth surfaces, such as trimmed ones. This is likely due to a lower accessibility of the peptides on those rough surfaces. These observations were confirmed by cell biological assays and demonstrated that a higher accessibility of the peptides for integrins expressed by osteoblasts correlated well with the enhancement of cell numbers.

This assumption is in line with a recent work of Pegueroles et al.[184] They studied the adsorption of FN to titanium surfaces with varying degrees of roughness and subsequent osteoblast adhesion. It was observed that on rough topographies FN was adsorbed preferably on peaks of the surface and an improved initial cellular interaction was observed with increasing roughness. Consistent with this finding, FN fibrils produced by osteoblasts accumulated mostly on top of rough surfaces (topographic peaks) and did not reach the surface valleys which showed almost no FN matrix.[184]

We can summarize that the surface roughness in combination with peptide coating can have a synergetic effect on osteoblast adhesion. In our experiments osteoblasts prefer most sandblasted textures coated with peptide 1.[177]

2.4.5 BSA blocking of Ti6Al4V disks influences osteoblast adhesion

In this section we focus on the influence of BSA blocking in our set-up. When performing an ELISA test, after the coating with the testing compound, BSA or fat-free milk[185] are applied to block any uncoated sites of the surface. In *in vitro* systems the treatment with BSA is required to avoid unspecific binding, as the blocking molecules lead to a saturation of free binding sites on the corresponding surfaces. This saturation is achieved by filling the surface with an almost totally covered layer of molecules. To investigate in particular the specific influence of the RGD peptides a blocking step with BSA is necessary and useful. However, the implant material that will be used *in vivo* is not treated with any BSA. Therefore, it is reasonable to check also the set-up without BSA blocking.

The objective of the following experiment was to check the specific influence of peptide binding on osteoblast adhesion with and without the step of BSA blocking. The results are shown in Figure 40. In (1) we performed the blocking step before the application of the cells, whereas in (2) this procedure was skipped. Both times the hexosaminidase test was accomplished as described in the materials and methods section.

With BSA blocking the peptide coating increased the cell attachment efficiency of osteoblasts by a factor of 1.7 on trimmed and on matt finished, and 1.2 on sandblasted surfaces, respectively. On all surfaces the coated and uncoated disks differed in osteoblast attachment level and the coating with peptide lead to more adhered osteoblasts.

When this experiment is performed without accomplishing the blocking step, all surfaces independently from their treatment with or without peptide provided similar numbers of adhered cells. No effect of peptide coating was observable; regardless of the surface roughness, cell attachment efficiency did not differ significantly. Even between the different uncoated surfaces the cell number remained comparable, meaning that cells adhere on all surfaces equally strong.

Coating the surface with BSA blocks the attachment of all kind of cellular receptors. The BSA molecule has a cigar shape and is 140 Å in length and 40 Å in height.[186] On surfaces coated with RGD peptides and blocked with BSA the only possible binding sites for cell receptors are the RGD peptides that are spread over the area and serve as binding islands for cell attachment.

We have shown before that a minimal distance of around 3.5 nm between the RGD motive and the surface is necessary for integrin binding.[173] One explanation for this is the correct presentation of the RGD motif, so the head of the integrin is able to bind to its ligand. However, this requirement can also be based on to the presence of BSA on the surface. If all free sides on the surfaces are blocked with a BSA layer that is 4 nm thick, the peptides are supposed to be at least the same height. The RGD motif has to be presented towards the integrins on the cell membrane in a way that the binding site of the integrin head can interact with the peptide.

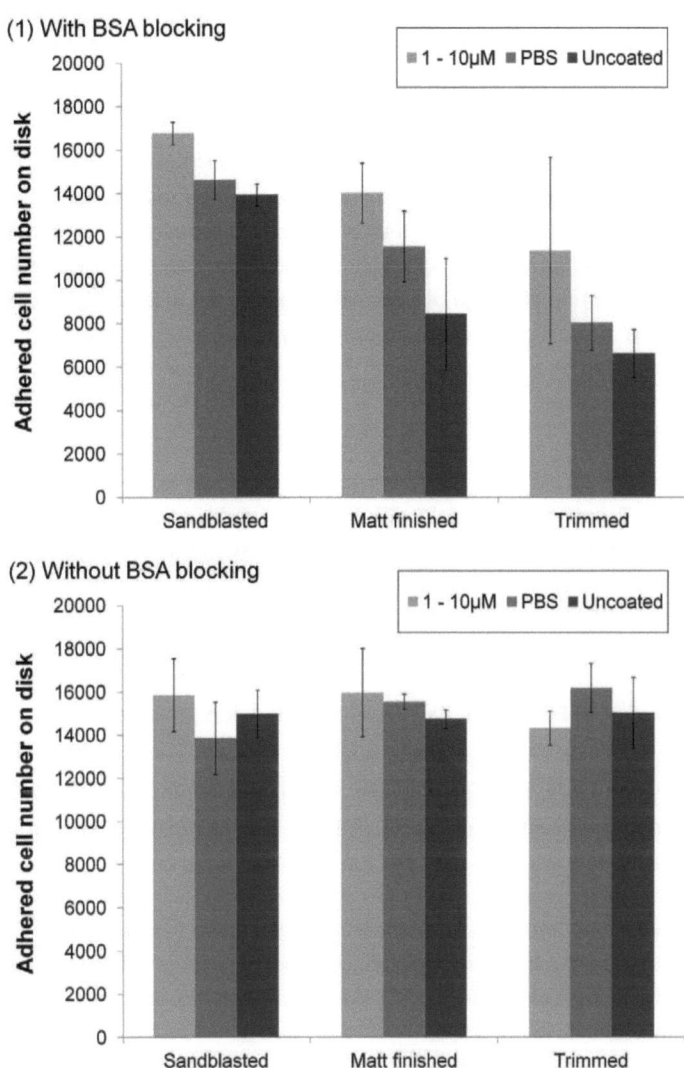

Figure 40: Influence of BSA blocking on the adhesion of osteoblasts (1) Ti6Al4V disks with different surfaces (trimmed, matt finished, sandblasted) treated with peptide 1 (10 µM in the coating solution) and PBS as well as uncoated disks were blocked with 1% BSA for 2 hour at 37 °C. The number of adherent cells from the osteoblast pool at passage 5 was estimated after 3 h of incubation by the hexosaminidase test. (2) The analog set-up was used, but the BSA blocking step was skipped.

On peptide coated surfaces blocked with BSA, the binding sites for cells are limited and cells can only bind to RGD peptides. If the surface is coated with RGD peptides, but the "free" spots are not blocked with BSA, cells will also get connected to the oxide layer of Ti6Al4V.

Not only the positive effect of peptide coating is unverifiable on surfaces that were not blocked with BSA, but also the influence of surface roughness is negligible. In general osteoblasts adhere on Ti6Al4V surfaces independently of the roughness, caused by the affirmatively interaction between the cells and the oxide layer of Ti6Al4V. Without the blockage of BSA, cell receptors are also able to bind to titanium alloy surface or more precisely to the oxide layer that is covered with different ions, peptides and proteins from the media. Although in stress media less peptides and proteins are present, the surface is immediately covered with low-molecular substances.

In summary the blocking with BSA results in a surface on which cells are only able to bind to RGD peptides. In these situations the roughness of the disks has great influence on cell adhesion and cells prefer rough surfaces to adhere. In absence of BSA cells can bind to the oxide layer of Ti6Al4V disks and no dependency on the roughness or RGD coating can be observed.

2.4.6 Amount of FCS in media affects the adhesion process

In a next step we investigated the influence of fetal calve serum (FCS) on the adhesion process on disks coated with RGD peptides. The amount of FCS in cell culturing media varies typically from 0 % to 20 % in literature, depending on the test system performed, the cells used and the definition of the project.[97,165] In contrast to BSA, that is never present in the *in situ* situation, analogous components of FCS are present, when an implant gets inserted into the human body. During surgery the implanted material will get immediately covered by a lot of different molecules and proteins from the blood serum.[73,78] For example, fibronectin and vitronectin are found in abundance in the blood.[187,188]

To mimic the natural case *in vivo*, we performed the assay adding FCS to the media. In order to obtain a better understanding of the interaction between osteoblast and the coated surface we additionally perform the experiments in stress media (without any FCS) and neglect the effect of serum. It is also worth mentioning that cells suffer

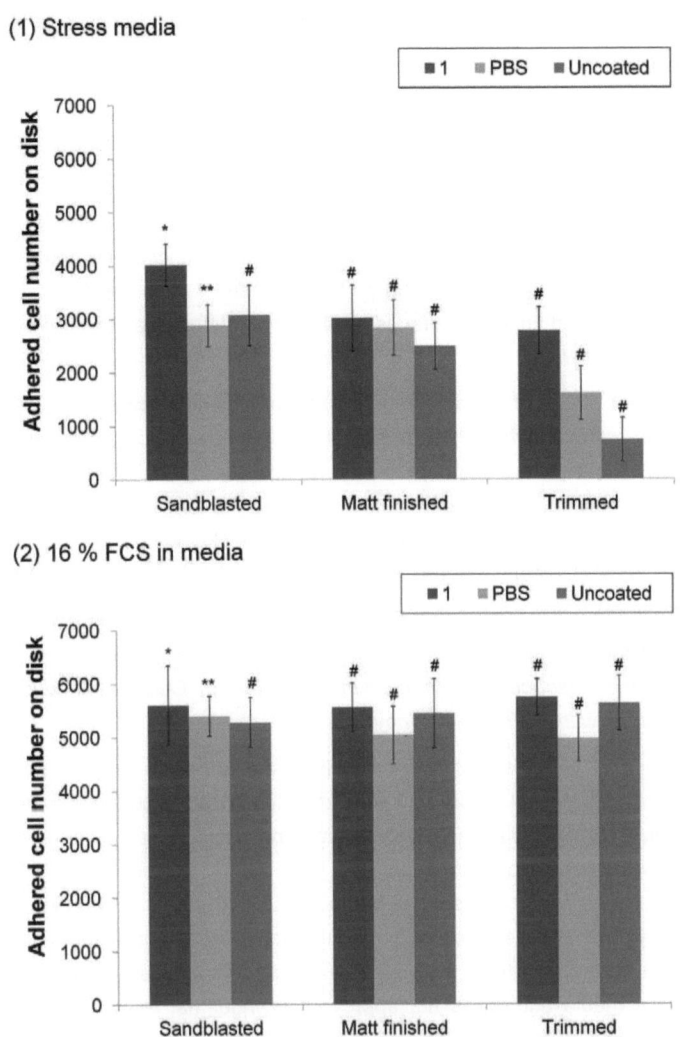

Figure 41: Influence of FCS amount in media on the adhesion of osteoblasts (1) Ti6Al4V disks with different surfaces coated with Peptide 1 (100 μM in coating solution) and PBS as well as uncoated disks were used. Osteoblast pool (passage 6) was suspended in stress media and the number of adherent cells was estimated after 3 h of incubation by the hexosaminidase test. (2) The analog set-up was used, but the media was accumulated with 16 % FCS. Between (1) and (2) the p values for the corresponding disks were calculated with the TTest: * = p≤ 0.05, ** = p≤ 0.01, and # = p≤ 0.001.

after a distinct time cultured in stress media without FCS. We monitored osteoblasts to change their morphology after around 10 h in stress media.

We performed the hexosaminidase assay on RGD peptide coated Ti6Al4V disks with 16 % FCS and without FCS (stress media) in Dulbeccos modified eagle Medium (DMEM) media, all other parameters remained constant and the protocol is described in the materials and methods section.

In Figure 41 we show the results. In comparison to the uncoated control osteoblast adhesion increased by a factor of 3.8 on trimmed, and 1.3 on sandblasted surfaces, respectively. On matt finished surfaces no significant difference between the coatings applied was measured.

The addition of FCS to the media changes the result significantly for all disks. On all surfaces independently from their treatment with peptide or not, similar cell numbers and therefore cell attachment efficiency is observed. The influence of surface roughness is minimized and sandblasted surface is no longer the preferred surface. This is in agreement with the results of Duewelhenke et al. who found also no differences on analogical surfaces in media with 20 % FCS.[189]

We can summarize that when FCS is added to the media both the effect of peptide coating and the influence of surface roughness disappeared similar to the result obtained in the absence of BSA blocking (see 2.4.5). Osteoblasts adhere on all surface textures with the same potency.

The presence of proteins that support cells in the adhesion process or other cell relevant activities is an important topic for tissue engineering. In absence of alternatives, FCS as serum additive is commonly used in cell culture. The serum provides a wide variety of macromolecular proteins, low molecular weight nutrients, carrier proteins for water – insoluble components and other compounds necessary for *in vitro* growth of cells, such as hormones and attachment factors. However, as the composition is not defined, by good manufacturing practices rules the application of FCS as a media additive is not approved for industrial tissue engineering. The bivalent situation about the use of FCS in adhesion assays is addressed by Siebers[73] and Duewelhenke.[189] When there is less amount of serum in the applied media, also a lower amount of proteins could adsorb on the substrate. However, the process of protein adsorption is highly dynamic and the composition of bound proteins on the surface changes rapidly, as they can detach again, are replaced by other proteins, get enzymatically degraded, denatured or undergo conformational changes.[37,73,190,191] In contrast to this, the coated peptides are permanently immobilized on the surfaces; moreover, these molecules are much smaller than proteins and can interact with integrins and therefore attract cell adhesion.

At this point, we would like to comment on some general aspects for *in vitro* studies. Per definition *in vitro* assays are always performed in surroundings that are non-physiological. Depending on the given scientific question only single parts of the complex *in situ* set-up are chosen to study one component in detail via an *in vitro* study. This must be kept in mind whenever interpreting data obtained by different *in vitro* models. Related to the application of RGD peptides, this aspect is discussed by Bellis et al.[192] For example the fact that the RGD domain will not react in isolation as well as the lack of serum in the media prevents a translation into the *in vivo* situation.[189,192] Duewelhenke et al. brought up the idea that in a wider sense the race for the surface is determined by proteins, peptides and small molecules, and not by cells itself.[189] Within seconds the surface of an implant material is covered by all kind of molecules present in blood serum and cells hardly have any direct contact to the oxide layer of a metal implant.[191] Another important factor for the comparison of different experiments is the level of standardization, e.g. for materials, cells and techniques that are applied. In the field of RGD peptides many varying scientific questions are addressed and therefore a broad variety of surface structures, implant materials, used cells, and methods for measuring cell adhesion is published. Ponche et al. and also Siebers et al. stress the lack of comparison in literature because information about surface roughness and organization is not documented and/or no standards for the documentation are defined.[73,159] Characteristic data differ according to the testing system applied. It is important to consider relevant structures, both in nanoscale and micro-scale.[159] Furthermore, the adhesion process is highly dynamic and time relevant. Depending on the chosen time point, the results might vary, as aspects like the adsorption of peptides and proteins from the media, the integrin expression pattern of osteoblasts, the mixture of cells present on the implanted surface and so forth are not fully understood.

We reviewed the adhesion process in detail in section 1.2.2 and stressed the importance of integrins in cell adhesion. Coating the surfaces with RGD peptides improved adhesion of different cells both *in vitro*[83-85] and *in vivo*.[86] Moreover, we showed that cells also adhere on surfaces without peptide coating. Osteoblasts interact with the oxide layer of titanium and its alloys in a non-specific way and all other present cells behave similarly. The coating of RGD peptides on the implant material can speed up the attachment process and moreover these molecules are able to regulate the preferred cell type due to their integrin selectivity. We can summarize that the use of peptides coated on the surface benefits cell adhesion.

2.4.7 SEM images show spreading of cells in detail

We studied the morphology of osteoblasts on surfaces coated with peptide **1**, peptide **2** or uncoated with a SEM. The images of the adhered osteoblasts are shown in Figure 42, Figure 43, and Figure 44

On trimmed disks osteoblasts appear flattened and are aligned following the circular grooving of the disks (Figure 42). The morphology of the cells cannot be distinguished on disks coated with peptide **1** or **2** or on disks treated with PBS. Osteoblasts show a similar shape on trimmed surfaces when compared to cell culturing plastic.

For matt finished textures osteoblasts do not look much spread and also the alignment is not maintained on these surfaces (Figure 43). Cells span randomly and are smaller in size compared to the other surfaces. We find differences between the surface treatments on matt finished disks. The disks coated with peptide 1 or treated with PBS present a better surrounding for osteoblasts. On both surfaces the cells adhere almost confluent, whereas on matt finished surfaces coated with peptide **2** the appearance of osteoblasts is changed. Cells adhere isolated and the edges look bulging.

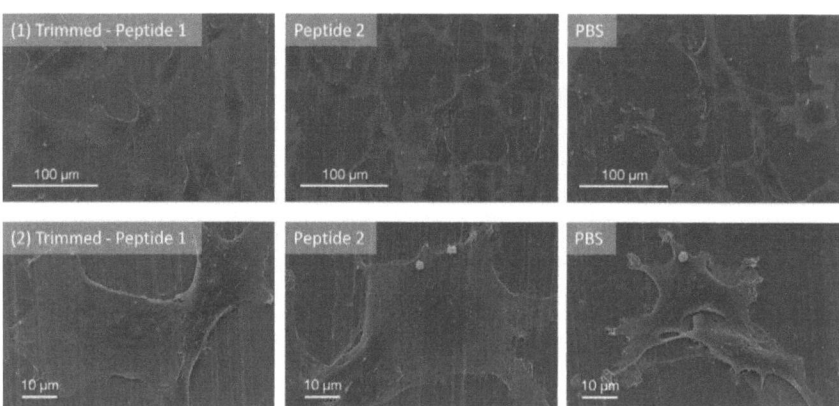

Figure 42: SEM images of adhered osteoblasts after 3 h of incubation time on trimmed Ti6Al4V disks (1) Coated with peptide 1 or 2 or treated with PBS (2) Same disks in higher magnification.

On sandblasted surfaces, cells show a cuboidal shape with a large number and long filopodia (Figure 44). The cells spread over valleys bridging the rough surface. The filopodial extensions are strongly developed and outstretched, this is a common morphology found in surfaces with increased roughness.[100,165,172] On sandblasted surfaces we do not find strong differences between the RGD peptide coated or PBS treated disks.

Cell morphologies strongly depended on the topography of each surface and were not influenced by the presence or absence of peptide coating. On smooth surfaces cells present an ECM that is stronger organized with focal contacts regularly distributed all over the surface. On the contrary, in a rough environment focal contacts are established mainly at the edges of the cells, where the cells are in contact with the material.[76] For this reason, on smooth surfaces cells seemed to be more spread and extremely flattened. In contrast, on rough substrates cells do not need extensive spreading to achieve focal contacts and hence showed cuboidal geometries.[76,100] The cell shape is influenced by roughness organization in the micrometer size. At this scale, the bound peptides do not have major influence,[158] as the RGD peptides range in nanometer dimension.

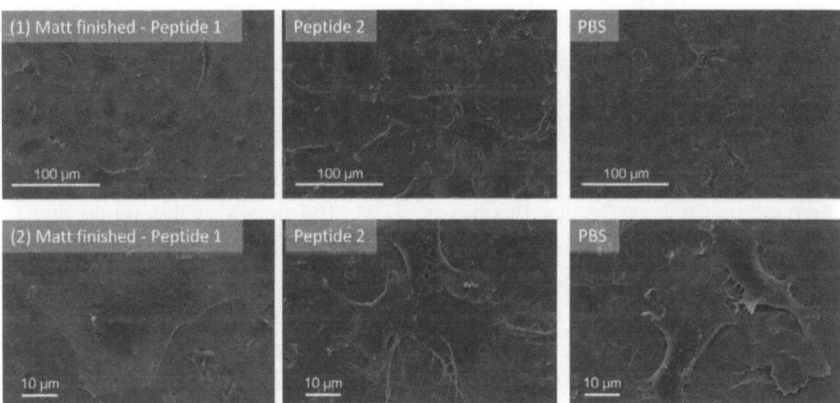

Figure 43: SEM images of adhered osteoblasts after 3 h of incubation time on matt finished Ti6Al4V disks (1) Coated with peptide 1 or 2 or treated with PBS (2) Same disks in higher magnification.

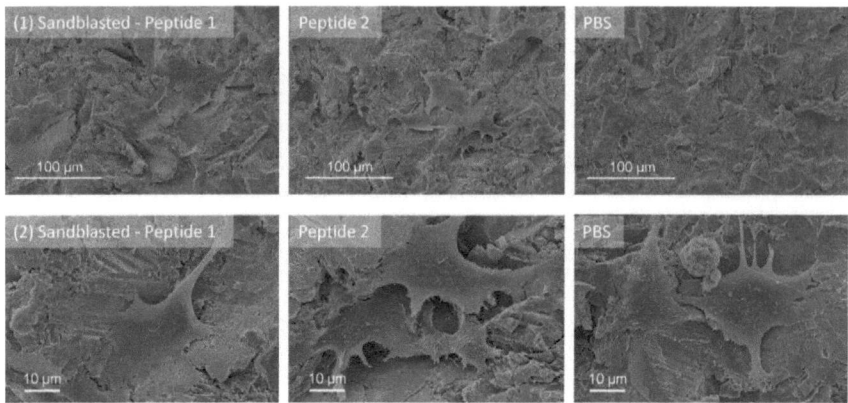

Figure 44: SEM images of adhered osteoblasts after 3 h of incubation time on sandblasted Ti6Al4V disks (1) Coated with peptide 1 or 2 or treated with PBS (2) Same disks in higher magnification.

2.4.8 Tracking of spreading shows effect of peptide and FCS on glass surface

The following investigations show the continuous spreading of osteoblasts over a time period of 6 h on glass slides to analyze the progression of the adhesion process.

These experiments were performed in cooperation with Prof. Bausch and Dr. Fernandez, Chair of Cellular Biophysics, TUM. The details of the set-up and the production of the channels are described in the materials and methods section. The glass surface of the channels was coated with 100 µM peptide 1 and 2 overnight. After blocking with BSA, osteoblasts were inoculated in stress media or media containing 15 % FCS. The whole system was monitored over 6 hours by every 30 seconds taking a picture of different areas within the channels. The level of spreading was measured by an image recognition program; we tracked the silhouette of the cells and the program calculated the surface area (shown in Figure 45). The area of the first measured time point was set to 100 % and we normalized the surfaces determined at later points to that surface. On three independent channels three cells were analyzed. The standard deviation is given as percentage of the calculated standard deviation for all analyzed cells.

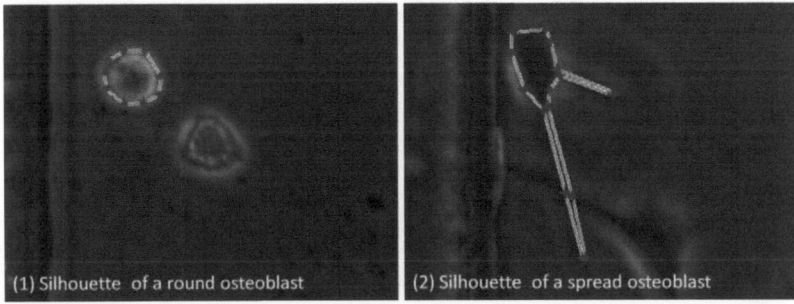

Figure 45: Silhouette of a round and spread osteoblast. (1) Cell shape directly after the inoculation at 0 h (2) Cell shape of the same cell after 6 h of incubation.

Figure 46 illustrates the influence of peptide coating in stress media. On uncoated surface there is little enhancement in spreading after 1 h, but osteoblasts kept the contact area to the surface constant. Peptide 1 and 2 coating on the surface influences the spreading in a positive way; already after 2 h a raise by the factor 3 for peptide 2 and 2.5 for peptide 1 is determined. The spreading takes place very fast and the peptide coating benefits this process.

Besides the coating, also the effect of media was also investigated. In Figure 47 we present the results obtained with stress media or media with 15 % FCS on uncoated surface. Without any FCS in the media the spreading area remains at a slightly increased level compared to the spreading detected at the beginning of the measurement (t= 0 h). Adding 15 % of FCS in the media boosts spreading by a factor of 4 after 4 hours and the spreading remains constant until the end of the measurement.

The proteins and additives of FCS have a positive effect on the spreading behavior of the cells. 15 % FCS in the media is enough to create a surrounding for the osteoblasts in which they adhere and spread. In the presence of serum a mixture of proteins adsorbs and gets sorted on the oxide layer of Ti6Al4V disks.

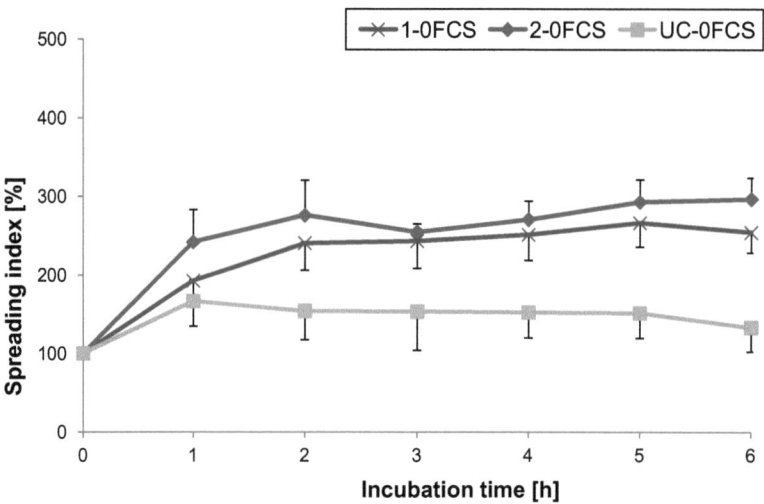

Figure 46: Level of spread cells in dependency of the peptide coating, detected over 6 hours in stress media (0FCS = DMEM without FCS); The glass surface was coated with 100 µM of peptide 1 (1), peptide 2 (2), or no treatment was performed (UC).

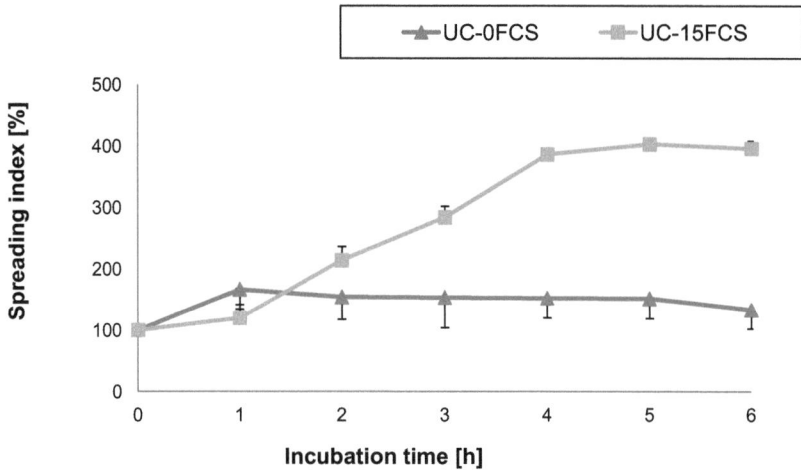

Figure 47: Level of spread cells on uncoated surface in dependency of the FCS level in media detected over 6 hours; The glass surface was uncoated (UC) and once no FCS (0FCS) was present in the media, whereas the other time 15 % FCS (15FCS) was added to the media.

In stress media cells are exposed to a different situation, as reviewed in section 2.4.6. Certainly the surface is also covered by ions and small molecules, but this situation is clearly different to surfaces that are in contact with media containing 15 % FCS. Therefore in stress media osteoblasts interact with the BSA blocked surface and cannot spread in a natural way. As discussed previous in chapter 2.4.5 the presence of BSA is artificial and non physiological for osteoblasts.

The results obtained on surfaces coated with peptide **1** and in media with or without FCS are shown in Figure 48. When coated with peptide **1** without FCS in media the level of spreading increases up to 200 % of the original value after only 1 h, the cells were spread only little more after 6 h (about 250 %). With 15 % of FCS in the media, the development of spreading differs over the experimental time: The presence of FCS decelerates the increase of spreading, after 3 h the cells spread to the same extent than the probes without FCS, however, after 6 h the spreading area is around 300 % of the original value and therefore slightly higher than that without FCS.

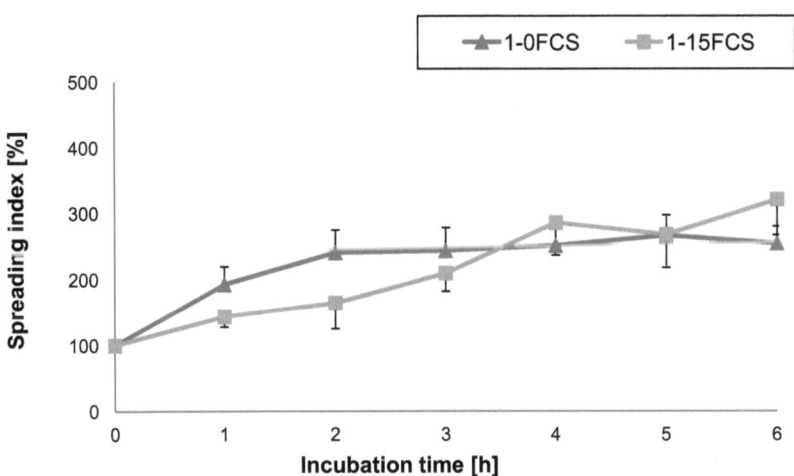

Figure 48: Level of spread cells on surface coated with peptide 1 (1) in dependency of the FCS level in the media detected over 6 hours; Once no FCS (0FCS) was present in the media, whereas the other time 15 % FCS (15FCS) was added to the media.

The presence of FCS in the media highly influences the spreading process on surfaces not coated with peptide. Osteoblasts in media without FCS adhere to that surface, but the spreading index does not increase over the duration of the experiment. Another important factor is the presence of peptides on the surface. We showed that peptide coating fastens the spreading process, predominantly in stress media. Within the first hours on peptide coated surfaces cells spread more when cultured in FCS-free media. In contrast to our previous experiments this test was performed on a glass surface.

We can conclude that on this surface the peptide binding and the presence of FCS do not show a synergetic effect. We find the highest spreading indices on uncoated channels with 15 % FCS in the media. It is important to note here, that after around ten hours the cells start to suffer in stress media. However, in media containing FCS osteoblasts can grow properly. Cells are sensitive to their surrounding and react in vitro clearly on the presence of coated peptides and media additives.

Degasne et al. investigated the adhesion in presence of FCS, fibronectin and vitronectin and found a correlation between the media additives and the spreading. They found that in stress media (without FCS) cells are packed and spread less than in media containing an additive that supports the adhesion process.[165] Our results are in agreement with these findings.

2.4.9 Adhesion process is highly time dependent

In this section we focus on the time dependency of cell adhesion. The effect of time is investigated by different methods: first the spreading was analyzed by fluorescent staining after 1 h and 3 h. The second approach was accomplished by the hexosaminidase test after 3 h and 10 h (this test is not suitable to investigate the very early stage of adhesion). In all experiments the RGD coating boosted cell adhesion in the early phase of adhesion. After a distinct time period dependent on the particular test, the accelerating effect of RGD coating is no longer detectable, meaning all cells have adhered equally strong on the different surfaces and the saturation regime started.

Influence of RGD coating is critical at the early adhesion process

To achieve a better understanding of the time dependency on the adhesion process, we monitored cell adhesion and spreading after 1 h and 3 h of incubation by means of fluorescent microscopy. We obtained a direct insight in the spreading process of cells, by observing the osteoblasts through the microscope.

In Figure 49 we show the images of adhered osteoblasts on PBS treated trimmed disks after both incubation times. After 1 h (left) osteoblasts still exhibited a round shape, whereas after 3 h (right) almost all of them possessed spread morphology and started clustering together.

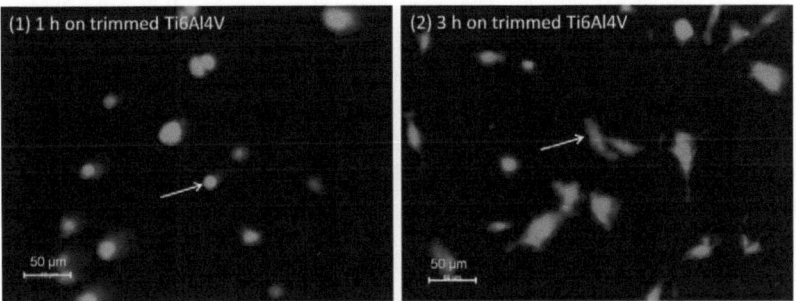

Figure 49: Images of adhered osteoblasts (passage 4) after 1 h (1) and 3 h (2) on trimmed Ti6Al4V disks treated with PBS. Cells were fluorescently stained with FDA for 10 min. The arrows indicate a 'round' cell in (1) and a 'spread' cell in (2).

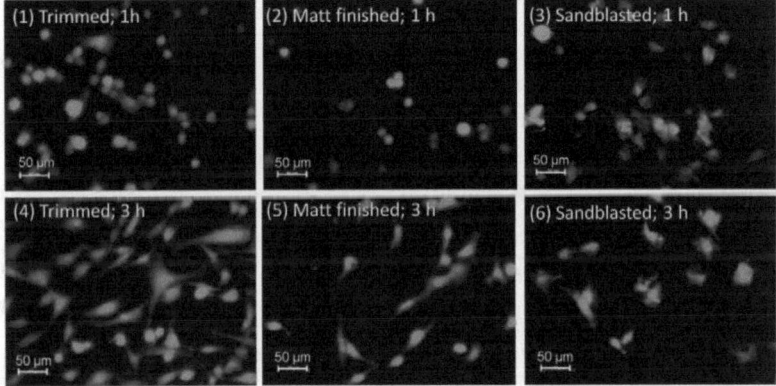

Figure 50: Images of adhered osteoblasts (passage 4) after 1 h and 3 h on Ti6Al4V disks coated with Peptide 1. After 1 h incubation time (1) trimmed, (2) matt finished, (3) sandblasted and after 3 h incubation time (4) trimmed, (5) matt finished, (6) sandblasted.

The images of adhered osteoblasts on Ti6Al4V disks with different surface textures are illustrated in Figure 50. On trimmed and matt finished surfaces cells still exhibited a round shape after 1 h, whereas on sandblasted most cells spread already onto the rough surface. After 3 h, independent of the surface, osteoblasts have adhered and showed the typical spread morphology. On trimmed and matt finished surfaces a minority of cells remained in a round shape.

Both, the peptide coating as well as the surface roughness clearly affected cellular spreading behavior (Figure 49, Figure 50, and Table 5). After 1 h, the use of peptides (**1** and **2**, respectively) supported osteoblast adhesion: in particular on sandblasted surfaces the number of adherent cells increased by around 20-fold (peptide **1**) and 16-fold (peptide **2**) when compared to non-coated surfaces. The same trend, however, to a smaller extent was observed for matt finished and trimmed disks: cell spreading increased by 8-fold (peptide **1**) and 4-fold (peptide **2**) on matt finished surfaces and by 7-fold (peptide **1**) and 4-fold (peptide **2**) on trimmed surfaces. After 3 h, the effect of peptides on cell adhesion and spreading was less pronounced. Regarding sandblasted and trimmed surfaces osteoblasts spreading reached over 90 %, regardless of the conditions used.

Table 5: Percentage of spread osteoblasts in correlation with the total number of cells applied on coated Ti6Al4V disks.[a]

Surface	Time	PBS (%)	Peptide 1 (%)	Peptide 2 (%)
Sandblasted	1 h	4.6±1.5	97.4±1.1	76.7±1.0
	3 h	96.0±3.2	100.0±2.7	95.2±1.7
Matt finished	1 h	4.7±7.6	35.3±6.1	20.4±6.8
	3 h	60.0±4.7	94.4±8.6	67.6±3.9
Trimmed	1 h	10.5±3.6	76.1±0.5	40.5±2.3
	3 h	91.6±3.3	96.1±1.5	94.3±3.2

[a] Ti6Al4V disks with different surfaces (trimmed, matt finished, sandblasted) were coated with peptide 1 or peptide 2 (100 µM), respectively, overnight. Pooled osteoblasts were used at passage 4 and incubated for 1 h or 3 h.

The majority of in vitro investigations from cell-implant interactions focus either on cell attachment after several hours or on proliferation during several days of culture.[98] In our studies, we concentrated on the early adhesion process and not on the proliferation rate. It has been reported that surface properties influence the interaction of cells with the implant material at an early stage[183] and we considered that the influence of peptide coating might be more crucial within the first hours of cell adhesion.[158]

In agreement with that, our results indicate that coating with RGD peptides was critical during the first hours of the adhesion process. When measuring cell spreading, the strongest effects of peptide coating were observed after 1 h. Cells exposed for 3 h did not show a difference in their spreading pattern no matter whether peptides were present or not.[177]

Analysis of cell numbers emphasizes time dependency of adhesion process

The second approach for investigation the adhesion process in terms of time dependency was performed with the hexosaminidase test. In this test we analyzed the adhesion situation after 3 h and, additionally, after 10 h incubation time. Another parameter remained constant to the previous experiments. Ti6Al4V disks (trimmed, matt finished or sandblasted) were coated with peptides **1** or **2**, respectively, for detailed description see the materials and methods section.

We show the results of these experiments in Figure 51. After 3 h adhesion time, the peptide coating as well as the surface roughness had a clear effect on the adhesion process. We already discussed this situation in chapter 2.4.4 in detail.

In contrast to this, after 10 h incubation time, cell numbers are very similar around 16,000 on all disks independently of the peptide coating or the surface properties. Small differences occur on trimmed surface, but no significant variations were identified.

One possible explanation for these findings is the fact that an incubation time of 10 h is too long to investigate the effect of roughness or the influence of peptides on the adhesion process. During the long incubation of 10 h, osteoblasts have enough time to adhere even on the less preferred surfaces, e.g. on trimmed. Within this time period they are able to produce their one ECM and build a network with other cells. The acceleration benefit through peptides coated on the surface takes place during the very first part of the adhesion and this point in time (10 h) is too late to investigate those issues. As the hexosaminidase test is an end point measurement we cannot state anything about the shape or the condition of the cells, but only about the number of adhered cells.

In summary, the effect of peptide coating or roughness parameters can be demonstrated by the hexosaminidase test only at certain time points, for example after an incubation period of 3 h. If this incubation time is extended up to 10 h, these differences are not detected anymore. This agrees with our results obtained by fluorescent staining, where the results demonstrated that cell spreading was influenced by peptide coating and surface roughness only within the first hour.

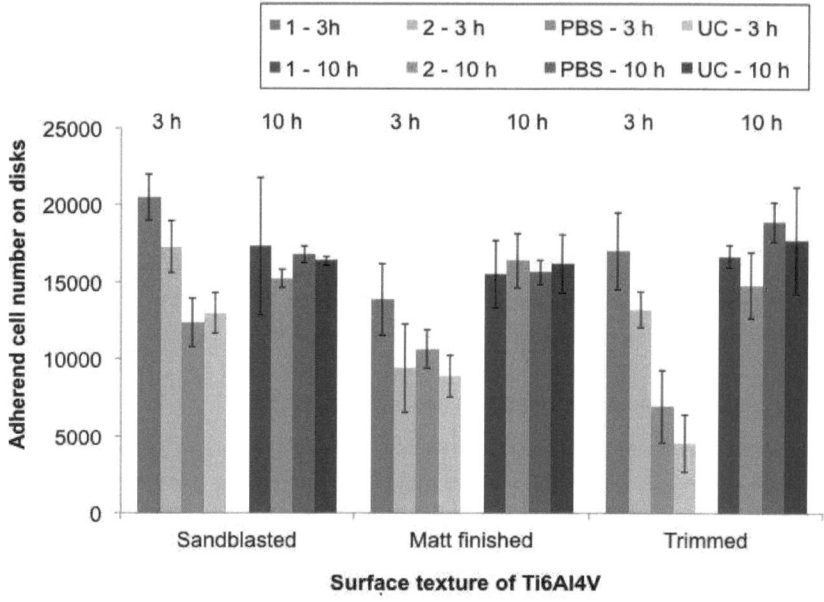

Figure 51: Cell numbers on different Ti6Al4V disks coated with peptide 1 and 2 after 3 h and 10 h, disks treated with PBS and uncoated were used as control. The osteoblast pool was suspended in stress media (DMEM without FCS) and the number of adherent cells was estimated after 3 h and 10 h of incubation by hexosaminidase test, respectively.

Depending on the chosen test, cells seem to vary in their adhesion potential. This cannot be explained with biological potential, because autonomously of the assay the adhesion process is performed following the same pattern. Therefore, the differences must originate in the performed tests. Indeed the treatment of cells differs within the assays, as both methods focus on different criteria. The fluorescent staining is based on a direct visualization of the adhesion process, so all cells are examined and no washing step is applied. The hexosaminidase test is detecting a metabolite and only adherent cells are supposed to be measured to obtain the correct cell number of adherent cells. Therefore, the protocol includes a rough washing step, determining to remove all non adherent cells.

The effect of RGD peptides saturates after a certain time period

Both methods described above indicate a maximum level of cellular coverage on the allocated material after a distinct time period. Osteoblasts bind to Ti6Al4V by undergoing the different steps of the adhesion process, beginning with the first contact and the appearance of Van-der-Waals forces. Peptide coating increases the attractiveness of the surface, so more of these first spontaneous interfaces result in stable contact points and cellular coverage appears earlier. However, also without peptides bound on the surface, the oxide layer of the Ti6Al4V disks is attractive enough to cause cell adhesion. Cell attachment especially on surfaces coated with RGD peptides can be described with a sigmoidal curve.[57,110,193,194] The situation on peptide coated and uncoated surface is illustrated in Figure 52. At time point (1) cells have spread on the surface with RGD peptide and the maximum coverage has been reached, whereas on the surface without coating only few cells adhered and the coverage is little. For our experiments this setting presents the circumstances after 1 h for the fluorescent staining and 3 h for the hexosaminidase test, respectively. Here, the differences between the probes are detectable and the benefit of peptide coating is evident. The second time point (2) describes the situation at the later time (in our experimental tests 3 h and 10 h, respectively) at which the maximal cell coverage is achieved for both samples. In that case the effect of peptide coating is saturated.

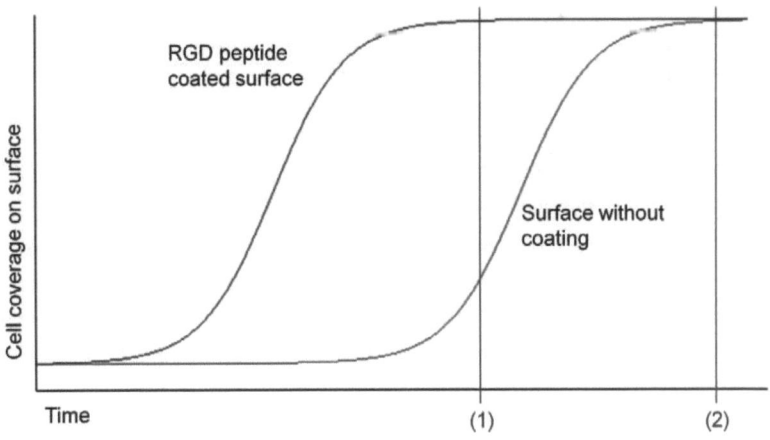

Figure 52: Schematic representation of cell coverage on different surfaces over the time. (1) and (2) mark two time points that describe diverse situations on the surfaces.

It is worth mentioning that this interpretation displays the *in vitro* situation with only one cell type attempting to adhere on the surface. *In vivo* all kinds of cell types are competing for the free spots on the implant material. In addition to bone cells, also fibroblasts and macrophages rival each other. As fibroblasts are said to be more active and faster adhering, the selective boost of osteoblast could help to avoid the building of fibrous tissue and by this ensure a stronger osseointegration. Recently Anselme et al. highlighted a comparable interpretation and concluded that the effect of coating is only for short term adhesion, but factors, like roughness influence the adhesion process also in the long term.[158]

We can state that after a certain time period, which depends on the applied number of cells, on the media used, and also on the assay performed, the entire surface of the used material will the covered by cells, provided that the material is biocompatible.

2.5 Surface structure is more important than surface treatments - demonstrated in a sheep experiment

In cooperation with CeramTec within the project "direct to bone" the new ceramic material "Biolox®*delta*" was tested. The aims of this project were to investigate the requirements of this new material in terms of medical surgery and conditions of manufacturing. Therefore aspects of design, material properties, and behavior *in situ*, e.g. the possibility of luxation were tested as well as cell toxicity and cell-material interaction *in vitro*. In a final step the most promising probes were implanted in Merino sheep for 4 and 12 weeks. We will discuss here mainly the results of the cell and animal testing of probes treated with RGD peptides. For a more detailed perspective see "Endbericht Verbundprojekt Direct To Bone".

PD Dr. Burgkart performed all animal surgeries, the push-out experiments were accomplished at the biomechanical department under supervision of PD Dr. Burgkart, and Susanne Kerschbaumer arranged the histological preparation and the staining. For the cell and animal experiments besides an unstructured (US) and a structured (S) probe, the following surface treatments were chosen: (1) Coating with hydroxyapatite (HA) as a common coating for bone implants, (2) plasma treatment (Plasma) to achieve a better surface wettability, and (3) one batch of probes was coated with the peptide RGDfK-Ahx-SH (RGD) in order to investigate the beneficial effect of the peptide on cell adhesion.

An overview of the disks is given in Figure 53. In the cell experiments, the disks coated with RGD peptide and treated with plasma showed after 24 h a higher proliferation rate than the other probes, however, after 96 h this effect was neutralized. As discussed previous in 2.4.9, the effect of surface treatment or peptide coating might not be detectable after a distinct time period, but might support human osteoblast at the race for the surface *in situ*.

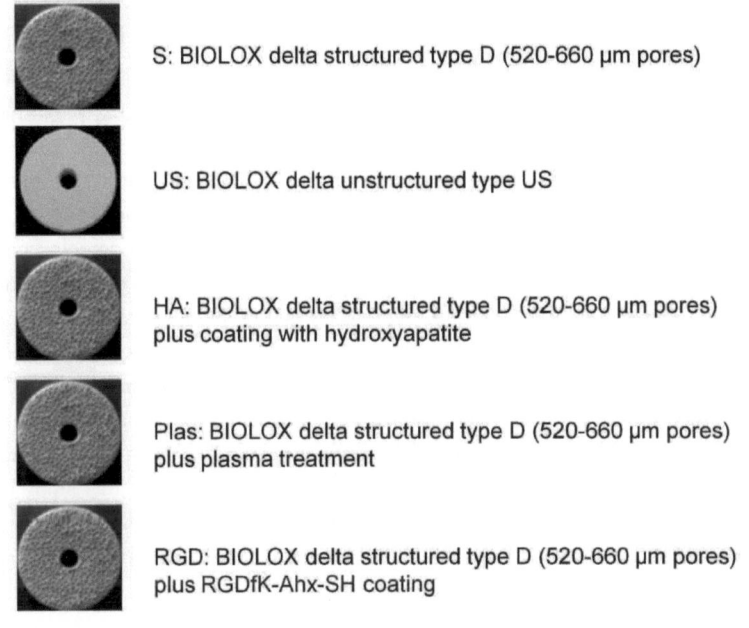

Figure 53: Overview of the disks with different surface structures and additional treatments used for cell experiments.

- S: BIOLOX delta structured type D (520-660 µm pores)
- US: BIOLOX delta unstructured type US
- HA: BIOLOX delta structured type D (520-660 µm pores) plus coating with hydroxyapatite
- Plas: BIOLOX delta structured type D (520-660 µm pores) plus plasma treatment
- RGD: BIOLOX delta structured type D (520-660 µm pores) plus RGDfK-Ahx-SH coating

Therefore all treatments were used for the animal study. The cylindrical probes, shown in Figure 54, were inserted in the distal femora and in the proximal tibiae laterally and medially. So in every sheep 4 probes per knee were implanted. The implantation time period was 4 or 12 weeks. After euthanasia of the sheep, the osseous integration was investigated by means of mechanical push-out tests (all probes from the distal femora) and histological staining (all probes from the tibia proximal). A detailed description can be found in the materials and methods section.

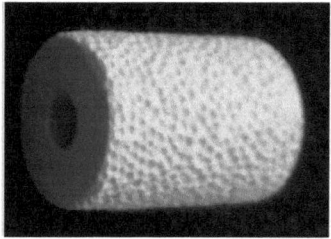 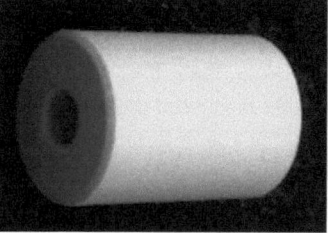

Figure 54: Structured and unstructured ceramic probes implanted in femora and tibiae.

Figure 55 illustrates the results of the push-out experiments. We can state, that the force needed to push-out the disks with a structured surface is around 7-fold higher compared to disks without structured surface. This tendency was found both after 4 and 12 weeks. Additional the surface treatment with plasma or the coating of probes with RGD or HA on structured surface did not enhance the force necessary to push-out the disks. Significant differences between the probes were detected neither after 4 nor after 12 weeks. The advantages of the different bioactive coatings, proven in the *in vitro* tests before, seem to be neutralized by the geometry surface structure. Furthermore, the significant progression of the necessary push-out force between the respective values for 4 and 12 weeks within one surface treatment is impressive and visualize the strong biomechanical interaction between bone tissue and implant. Although a direct comparison between different animal studies is not possible, as the animals and techniques are not identical, the tendency obtained in this study looks promising. For example after 4 weeks our values for structured ceramic probes are 6-fold higher compared to a study from 2005 on roughened titanium probes.[195]

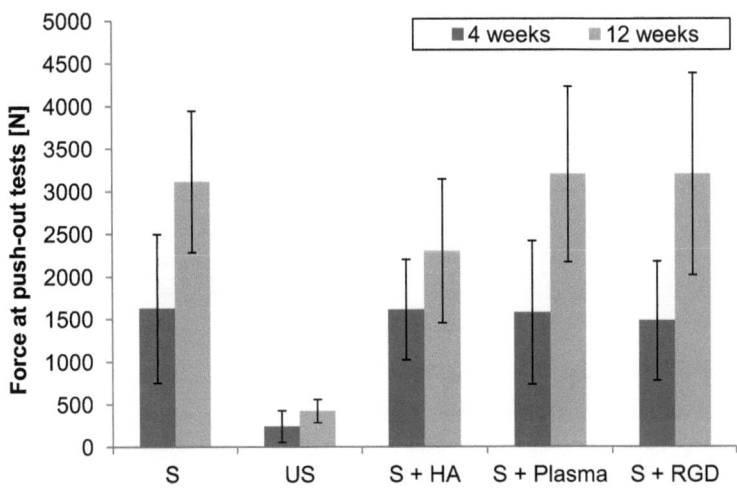

Figure 55: Push-out experiment for ceramic probes in sheep after 4 and 12 weeks. S = structured surface, US = unstructured surface, S+HA = structured surface coated with hydroxyapatite, S+Plasma = structured surface plasma treated, S+RGD = structured surface coated with RGDfK-Ahx-SH.

The histological staining delivered insight to the situation directly on the implant material. The cells in the tissue interacting with the ceramic were visualized and thereby also characterized for osteoblast. In Figure 56 and Figure 57 images of two samples implanted both in the identical sheep - the one in the right tibiae at lateral position, the other in the left tibiae at medial position, are illustrated. Figure 56 presents a structured ceramic probe that was coated with RGD peptides, whereas Figure 57 displays a structured ceramic probe without any treatment. In both samples a clear bony integration is detected, although not all parts of the implant were integrated in the same way and locally strong differences in the quality of bone ingrowth occurred. As cells grew in the holes of the structure and the tissue with cells counterfeited the structure of the ceramic, the implants look strongly anchored.

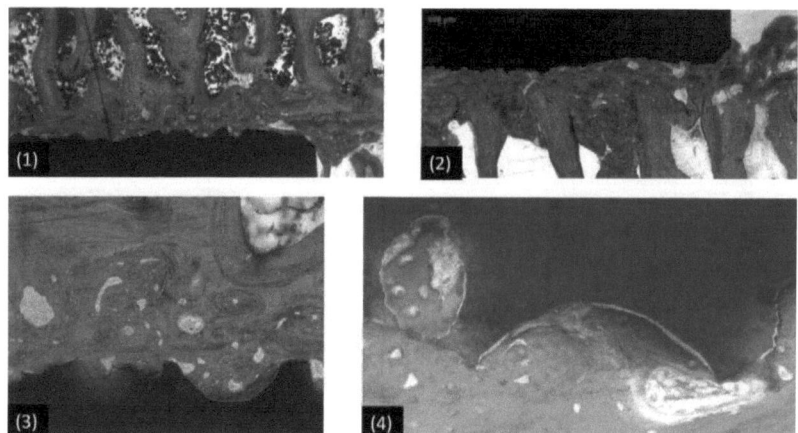

Figure 56: Example of a structured ceramic probe coated with RGD peptide. (1) Implant surface from the right tibiae at proximal lateral position with distinct integration of the implant material (magnification: 2.5 x) (2) Implant surface from the right tibiae at distal lateral position with marginal integration of the implant material (magnification: 2.5 x)   (3) Bony integration of the implant material (magnification: 10 x) (4) Bony integration of the implant material (magnification: 40 x).

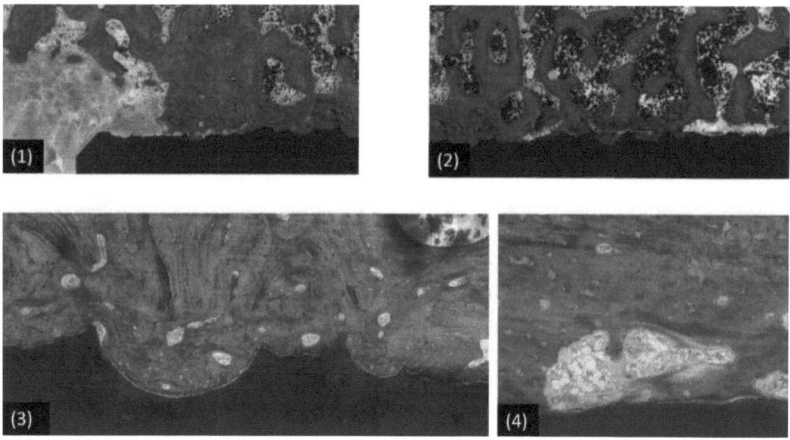

Figure 57: Example of a structured ceramic probe (1) Implant surface from the left tibiae at proximal medial position with local integration of the implant material (magnification: 2.5 x) (2) Implant surface from the left tibiae at medial position with marginal integration of the implant material (magnification: 2.5 x) (3) Completely bony integration of the implant material, the bone is mineralized (magnification: 10 x) (4) Poorly mineralization, on implant surface building of an lacunae, only fragmentary mineralization (magnification: 40 x).

To rank the results of the histological staining, all samples were evaluated in different categories as listed in Table 6. The classification was performed by means of a 4-step pattern ranging from "sporadically measured" to "profoundly measured".

Table 6: Counting results of the staining from cross sections of various probes.

4 weeks	US	S	HA	PLAS	RGD
Direct bone contact	3	2	2	2	3
Surface contact	2	2	1	2	3
Mineralization	2	4	4	4	3
Osteoblast activity	4	2	3	3	3
Bone vitality	4	2	2	3	3
Cell activity	2	3	2	3	2

[a] legend: 1 = sporadically measured; 2 = sparsely measured; 3 = moderately measured; 4 = profoundly measured.

12 weeks	US	S	HA	PLAS	RGD
Direct bone contact	3	3	3	4	3
Surface contact	3	3	3	3	4
Mineralization	4	4	4	4	4
Osteoblast activity	3	3	4	4	3
Bone vitality	3	4	4	4	4
Cell activity	4	4	4	4	4

[a] legend: 1 = sporadically measured; 2 = sparsely measured; 3 = moderately measured; 4 = profoundly measured.

For the RGD coated ceramic disks after 4 weeks good direct to bone contact, well established surface contact, mean mineralization and osteoblast activity, as well as moderate bone viability was detected. Only in terms of cell activity these probes came off with a reduced score. Compared to the other probes, they showed a constantly high integration performance in all categories. Structured probes in contrast demonstrated a profound mineralization, but there sparsely bone vitality, osteoblast activity, surface contact, and direct bone contact was measured.

After 12 weeks the situation looks similar. The probes with plasma treatment got the best evaluation in 5 out of 6 categories (only the surface contact was measured moderately), whereas disks coated with RGD received highest marks in 4 categories, including surface contact, mineralization, bone viability and cell activity. The other biofunctionalized surface, namely the probes coated with HA, performed similarly. For the structured and unstructured surfaces without any treatment or coating the values measured in the single categories were straight.

In conclusion this animal study demonstrated the importance of the implant surface. By changing an unstructured ceramic surface into a newly developed macroporous surface significant higher bone fixation was found. This effect continues from the 4 week group to the 12 week group. In contrast the biofunctionalization performed by plasma treatment, HA and RGD coating did not increase the fixation. By histological staining, the distinctive interaction between the tissue and the implant got observably and measurable. All probes performed well and only slightly differences within the different categories can be fund. In short, after 12 weeks the results of the histological staining identified by trend a better performance of the biofunctionalized probes, like plasma treatment, HA or RGD peptide coating.

3 SUMMARY

Despite numerous advances aiming to optimize material properties and biocompatibility as well as surgical tools for new generation implants, a fast, stable and long lasting osseointegration still represents a clinical challenge. To achieve this goal, a key aspect is the functionalization of implant surfaces in a way that they become both highly attractive and selective for osteoblast adhesion already starting in the early phase, i.e. after the insertion of the endoprosthesis.

In this thesis we have optimized the process of osteoblast adhesion on commonly used titanium alloys (Ti6Al4V) by the combination of two approaches: Biofunctionalization of the implant material by RGD peptide coating and modification of the surface roughness.

In a first step a pool of human primary osteoblasts was established and cells were characterized for osteoblast specific metabolites. For clear studies, a careful description of the applied Ti6Al4V disks was essential. By classifying the sample surfaces into different roughness categories, systematic tests on isotropic and anisotropic surfaces and additionally on varying levels of roughness became feasible. In integrin-binding studies it was shown that the functionalization of Ti6Al4V disks with RGD peptides significantly enhances $\alpha v \beta 3$-binding. It was demonstrated that in addition to the RGD integrin-binding motif, the other functional units of the peptide, namely the spacer and the anchor units, also play an important role. An optimal spacer-anchor system should guarantee a minimal distance between the surface and the RGD sequence. In our study thiol and tetraphosphonic acid anchoring groups were evaluated. Tetraphosphonic acids are known to be highly stable anchoring groups to titanium. This is corroborated in our experiments, where they showed an improved performance as compared to thiol anchors.

Experiments with primary osteoblasts yielded the coherent results. In toxicity tests it was demonstrated that all surfaces with and without peptide coating do not possess any toxic effect on cells. The chemical composition of an implant material has been described to influence the process of cell attachment and adhesion. In this context the native oxide layer of titanium surfaces, the presence of proteins, BSA and/or RGD peptides become highly important for both adhesion rate and strength of osteoblasts. Interestingly, the surface topography (e.g. surface roughness) is similarly important for cell adhesion as the functionalization of the surface. Some of these particular features on the surface can have synergistic effects and other factors might impair each other.

The highest increase on cell adhesion was found on smooth surfaces coated with RGD peptides. Most likely the accessibility of the peptides for cell-expressed integrins on these surfaces peptides accounts for this effect. In contrast, the accessibility of peptides that were adsorbed on more rough surfaces was lower, resulting in an attenuated effect of the peptides. Nevertheless, in absolute terms the combination of rough surface (S_a = 3.24 µm) with peptide coating yielded the highest adhesion rate. Therefore this presents the most promising alliance for clinical applications in further studies.

Furthermore, the treatment of surfaces with BSA was studied. BSA molecules occupied the free binding sites on the oxide layer of Ti6Al4V coated with RGD peptides. This results in a surface, where cells preferentially bind to RGD peptides, as all other possible binding sites are blocked by BSA molecules. Without any RGD binding sites on uncoated surfaces, the effect of surface roughness influences cell adhesion in a stronger way compared to RGD peptide coated disks. In absence of BSA blocking cells directly bind to the oxide layer of Ti6Al4V and no preference for roughness or RGD coating was observed in the performed tests.

A further, so far underestimated, parameter to be taken into account for the study of cell-surface-interactions is the composition of the cell culture media used (e.g. the presence and the concentration of FCS). The effect of RGD peptide coating on the adhesion process can be neutralized by the presence of other proteins, growth factors and blood or serum components. In experiments performed with media complemented with FCS no difference in osteoblast adhesion was found after 3 h incubation time, irrespective of coating and surface structure. It is worth mentioning that not only cells preferably bind to RGD peptide coated surfaces, but also integrin-binding proteins in the blood or serum. These bound proteins stimulate cells to adhere on the surface. The coating of the surface with peptides in an unspecific way increases the affinity towards both cells and proteins.

The kinetics of the adhesion process is another critical topic that has to be considered. We have demonstrated that the effect of coated RGD peptides is highly relevant within the first hours when the implant material becomes exposed to cells. The spreading of osteoblasts is significantly accelerated by the presence of RGD peptides immobilized on the Ti6Al4V disks. At long incubation times the influence of peptide coating is no longer observed. However, the benefit through accelerated cell adhesion within the first hour might be relevant for a stable and selective osteoblast-implant interphase.

In the animal experiments with sheep we affirmed the fact that the structure of the surface is highly important for the ingrowth of the entire implant. Both after 4 and 12 weeks the probes with a structured surface possessed a significant higher bone fixation. In addition, it can be stated that the biofunctionalization, for example coating with RGD peptides, improved the outcome of the histological staining after 12 weeks.

In summary we are aware that most results presented here are based on *in vitro* models of osteoblast adhesion and that *in vivo* adhesion is much more complex. After implant surgery, the implant material will rapidly be covered through the competitive adsorption of both proteins and cells. In the "race for the surface" not only cells, but also proteins play an important role. For a successful biointegration it is of utmost importance that osteoblasts bind to the implant at an early stage of this process, because only osteoblasts facilitate a stable osseointegration. Especially preference of osteoblasts over fibroblasts is crucial, as the latter build up an instable fibrous tissue around the implant material. Coating Ti6Al4V surfaces with RGD peptides effectively influences the osteoblast adhesion process at early stages. Furthermore, the combination of peptide coating with optimal surface properties has the potential to increase the clinical output of biomaterials in orthopedics and hence to be beneficial for numerous patients.

4 MATERIALS & METHODS

4.1 Materials

4.1.1 Chemicals

Registered trade name	Company	Reference number	City, Country
1-propanol	Merck	603-003-00-0	Darmstadt, Germany
3,3', 5,5''-tetramethylbenzidine (TMB)	Sigma	860336	Deisenhofen, Germany
4-Nitrophenyl-N-acetyl-beta-D-glucosaminide 98- 100 %)	Sigma	N9376-1G	St. Louis, USA
BSA (30 % Aqueous-solution)	Calbiochem/ Merck	126625	Darmstadt, Germany
$CaCl_2$	Merck	1.02378.500	Darmstadt, Germany
CD49e/ Integrin $α_5$ chain	BD Biosciences	555651	Franklin Lakes, USA
Ethylendiamintetra-acetat (EDTA)	Sigma	E2 628-2	St. Louis, USA
EZ-Link NHS-LC-LC-Biotin	Pierce	21343	Rockford, IL, USA,
Glutaraledehyde 25 % in H_2O (high purity)	Serva	23114.02	Heidelberg, Germany
H_2SO_4	Merck	1120801000	Darmstadt, Germany
KCl	Merck	1.04936.1000	Darmstadt, Germany

KH_2PO_4, pH 7.4	Merck	1.05099.1000	Darmstadt, Germany
$MgCl_2$	Merck	1.05833.1000	Darmstadt, Germany
$MnCl_2$	Merck	1.05927.1000	Darmstadt, Germany
Na_2HPO_4	VWR	28026.260	Leuven, Belgium
NaCl	Merck	1.06404.5000	Darmstadt, Germany
NaH_2PO_4	Merck	1.06580.0500	Darmstadt, Germany
NaOH	Merck	1.09137.1000	Darmstadt, Germany
NeutrAvidin-horseradish peroxidase conjugate	Sigma Life Science	P 6662-100 Tab	Deisenhofen, Germany
o-phenylenediamine dihydrochloride (OPD)	Sigma	P6662-50TAB	Deisenhofen, Germany
PFA	Merck	1.04006.1000	Darmstadt, Germany
primary antibody (CD49e)	Merck	1.09057.1000	Darmstadt, Germany
Rec Human Integrin alphaVbeta3 CF	R&D Systems	Bulk 3050-AV	Minneapolis, USA
SeramunBlau® fast Ready-to-use TMB/substrate solution for ELISA	Seramum Diagnostica GmbH	S-001-4-TMB	Heidesee, Germany
Tween 20	Roth	9127.1	Karlsruhe, Germany

4.1.2 Equipment

Registered trade name	Company	Reference number	City, Company
AxioCam ICc3	Zeiss		Oberkochen, Germany
BMG POLARstart Galaxy plate reade	BMG Labtech	P21009	Ortenberg, Germany
Cell filter (40 µm, Nylon)	BD Falcon	352340	Durham, USA
Counting chamber	Paul Marienfeld	06 400 10	Lauda-Königshofen, Germany
Critical-Point-Dryer CPC 030	BAL-TEC		Liechtenstein
Culturing disk (750 ml)	BD Biosciences	353112	Bedford, USA
Discovery.V8 SteREO	Zeiss		Oberkochen, Germany
Electronic cell counter (CASY)	Roche	2501126	Mannheim, Germany
Eppendorf Combitips plus (5 ml)	Eppendorf	0030 069.455	Hamburg, Germany
Eppendorf Easypet pipettor	Eppendorf	4421 000.013	Hamburg, Germany
Eppendorf Multipet plus	Eppendorf	4981 000.019	Hamburg, Germany
Eppendorf Reference	Eppendorf	4910 000.514	Hamburg, Germany

Falcon – Multiwell-plates (6, 24, 48, 96)	BD Biosciences	353046 (6) 353047 (24) 353078 (48) 353072 (96)	Bedford, USA
Falcon serological pipet (5 ml, 10 ml, 25 ml)	BD Biosciences	357543 (5ml) 357551 (10ml) 357525 (25ml)	Bedford, USA
Fluoroskan Ascent FL	Thermo	5210450	Waltham, USA
Incubator Hera Cell 150i	Thermo	51026280	Waltham, USA
Laboratory- Bench Class II Hera Safe	Thermo	HS12 51018097	Waltham, USA
Microscope Wilovert 30 Standard HF	Helmut Hund	008.0302.0	Wetzlar, Germany
Mikroskope Axiovert 200M	Zeiss		Oberkochen, Germany
Multiskan Ascent	Thermo	51118300	Waltham, USA
Nunc Multiwoll plates	Nunc A/S	150787	Roskilde, Denmark
Pipet tip	Biozym Scientific	721010	Oldendorf, Germany
Reaction tube (2 ml)	Biozym Scientific	710228	Oldendorf, Germany
Scanning electron microscope Leo 440i	Zeiss		Oberkochen, Germany
Sputter Coater SCD 005	BAL-TEC		Liechtenstein
Sterile filter (MILLEX GP, 0.22 µm)	Millipore	SLGP033RB	Carrigtwohill, Ireland

Tactile profilometer (MarTalk)	Mahr		Ingolstadt, Germany
Tube (15 ml, 50 ml)	BD Biosciences	352096 (15ml) 352070 (50ml)	Bedford, USA
Water bath SubAqua 14	Grant	1KO136010	Shepreth, Great Britain

4.1.3 Disks

Registered trade name	Company	Reference number	City, Company
Ti6Al4V (Diameter: 10 mm, Thickness: 2 mm)	Biomet	ASTM F136 & ISO 5832-3	Berlin, Germany

4.1.4 Cell culture and Immunochemistry

Registered trade name	Company	Reference number	City, Country
AEC+High Sensitivity Substrate Chromogen	Dako	K 3469	Glostrup, Denmark
Alexa Fluor®488 goat anti-mouse IgG	Life Technologies	A11001	Darmstadt, Germany
Alpha medium	Biochrom	F 0925	Berlin, Germany
Antibody Diluent	Dako	83022	Glostrup, Denmark
Anti-Collagen I (rabbit)	Quartett	2031500105	Berlin, Germany
Anti-Fibronectin (rabbit)	Dako Cytomation	A 0245	Glostrup, Denmark

Anti-Osteocalcin (rabbit)	Biotrend	97060-1515	Köln, Germany
Aqua.dest	Delta Select	PZN-8771079	Dreireich, Germany
Ascorbin acid	Sigma	A 4403	St.Louis, USA
Avidin/Biotin-Complex (Vectastain ABC Kit)	Vector Laboratories	PK 6100	Burlingame, Canada
Biotinylated secondary antibody Anti-Rabbit (1/ 200 in PBS)	Vector Laboratories	BA 1000	Burlingame, Canada
CD51 (αv)	Chemicon	MAB1956	Temecula, CA, USA
CD51/61 ($\alpha v \beta 3$)	Chemicon	CBL544	Temecula, CA, USA
CD61 ($\beta 3$)	Southern Biotech	9470-01	Birmingham, USA
Cell Freezing Medium-DMSO 1 x	Sigma	C6164	St. Louis, USA
Cytotocity Detection Kit (LDH)	Roche	11644793001	Mannheim, Germany
Dexamethasone	Sigma	D 8993	St.Louis, USA
DMEM (Dulbeccos modified eagle Medium)	Biochrom	FG 0415	Berlin, Germany
Dulbeccos PBS (w/o calcium and magnesium)	Biochrom	L 1825	Berlin, Germany

FBS Superior	Biochrom	S 0615	Berlin, Germany
Fluorescein diacetate (FDA)	Sigma	F7378	St. Louis, USA
Glycerol gelatin	Merck	1.09242.0100	Darmstadt, Germany
HEPES-Puffer (4-(2-hydroxyethyl)-1piperazineethanesulfonic acid)	Biochrom	L 1613	Berlin, Germany
Human fibroblasts cell line	Cell lining	1210411 or 1110411	Berlin, Germany
L-Glutamine	Biochrom	K 0283	Berlin, Germany
MEM Dulbecco w/o Ca^{2+}	Biochrom	F 9050	Berlin, Germany
MEM-Vitamine	Biochrom	K 0373	Berlin, Germany
NBT/BCIP Ready-to-use tablets (nitro blue tetrazolium chloride/5-bromo-4-chloro-3-indolylphosphate)	Roche	11697471001	Mannheim, Germany
PBS/Brij (0.01 %Brij in PBS)	Sigma	430 AG-6	St. Louis, USA
Pepsin	Sigma	P-7012	St. Louis, USA
Primocin	InvivoGen	ant-pm-2	San Diego, USA

Protein Block	Dako Cytomation	X 0909	Glostrup, Denmark
Proteinase K (1: 1000 in PBS)	Qiagen	19131	Düsseldorf, Germany
Trypsin/EDTA	Biochrom	L-2143	Berlin, Germany

4.2 Methods

4.2.1 Roughness measurements of Ti6Al4V disks

Classical roughness parameters were measured by using a tactile profilometer (MarTalk) on a randomly chosen surface area (16 mm^2) of the disks. For three times, every 10 µm one measurement was set. A series of roughness parameters were computed: the average roughness (S_a); the height difference between the highest and the lowest peak of the image (S_z); the root mean square (S_q); the asymmetry of the height distribution histogram also termed surface skewness (S_{sk}); the peakedness of the surface topography also known as surface kurtosis (S_{ku}); the largest peak height value (S_p); and the largest valley depth value (S_v).

4.2.2 Peptide coating of Ti6Al4V disks

Ti6Al4V disks were coated overnight at room temperature (RT) with 100 µL/disk of peptide solutions (100 µM) in PBS (10 mM Na_2HPO_4, 2 mM KH_2PO_4, pH 7.4, 137 mM NaCl, 2.7 mM KCl). Disks were subsequently washed three times with PBST buffer (PBS + 0.01 % (v/v) Tween 20) and blocked for 2 h at 30 °C with 100 µL/ disk of TSB-buffer (20 mM Tris-HCl, pH 7.5, 150 mM NaCl, 1 mM $CaCl_2$, 1 mM $MgCl_2$, 1 mM $MnCl_2$, 1% (w/ v) of BSA) for cell-free studies. In all cell assays we blocked the disks in 1 ml of 1 % BSA in PBS for 2 h at 37 °C.

4.2.3 αvβ3-adhesion assay

Coated disks were washed three times with PBST and incubated with 100 µL of 10 µg/ mL human biotinylated-αvβ3 integrin (Millipore) in TSB for 2 h at 30 °C. Biotinylation of the integrin was done in-house with sulfo-NHS-LC-LC-Biotin (Pierce, 20:1 molar ratio). After five washes in PBST, disks were treated with 100 µL/disk of 0.25 µg NeutrAvidin-horseradish peroxidase conjugate in TSB for 1 h at 30 °C. Finally, disks were washed five times and the binding was visualized by adding to each disk 100 μL of 1 mg of o-phenylenediamine dihydrochloride (OPD) dissolved in 2.5 mL of buffer (50 mM Na_2HPO_4, 24 mM sodium citrate, pH 5.0, 0.012 % (v/v) H_2O_2) for 10 to 15 min at RT. The reaction was stopped by adding 50 μL of 2 M H_2SO_4, and the absorbance was measured at 492 nm with a POLARstar Galaxy plate reader. All samples were analyzed in triplicate, and reproducibility was confirmed with at least three identical but independent assays. Control disks (uncoated or coated only with PBS) and a positive control (peptide 1) were included in all the assays.

4.2.4 α5β1-adhesion assay

Coated disks were washed three times with PBST and incubated with 100 μL of 10 μg/ mL human α5β1 integrin in TSB for 1 h at 30 °C. After three washes in PBST, disks were treated with 100 μL/disk of 1 μg primary antibody (CD49e) in TSB for 1 h at 30 °C. The disks were washed again three times in PBST, followed by adding 100 μL/ disk of 2 μg/ mL secondary antibody (anti-mouse IgG-POD) in TSB for 1 h at 30 °C. Finally, disks were washed three times and the binding was visualized by adding to each disk 100 μL of 3,3',5,5"-tetramethylbenzidine (TMB) for 3 to 5 min at RT. The reaction was stopped by adding 50 μL of 2 M H_2SO_4, and the absorbance was measured at 450 nm with a POLARstar Galaxy plate reader. All samples were analyzed in triplicate, and reproducibility was confirmed with at least three identical but independent assays. Control disks (uncoated or coated only with PBS) and a positive control were included in all the assays.

4.2.5 Obtaining and culturing primary human osteoblasts

To ensure the sterile conditions cell culture procedures were carried out in a microbiological cabinet (hood). Cells were grown in commercially available cell culture flasks or culture slides and incubated at 37 °C in a humid atmosphere containing 5 % CO_2. Cell culture medium used for adherent cells was DMEM (supplements according to the used cells see above).

Primary osteoblasts were isolated from cancellous bone obtained from surgical waste during total hip joint replacement surgery. The local Ethics Committee approved their use for scientific purposes (1307/05). The bone was cut into small pieces and incubated in calcium-free alpha-medium (Biochrom, Berlin, Germany) supplemented with 16 % fetal calf serum (FCS), 0.08 % MEM-vitamin (different amino acids), 16 mM 4-(2-hydroxyethyl)-1-piperazineethanesulfonic acid (HEPES), 1.6 mM L-glutamine, and priomycin at 37 °C and 5 % CO_2.[176] After 10 days, culture medium supplemented with 50 μg/ mL ascorbic acid and 10 nM dexamethasone was added in order to stimulate the osteoblastic phenotype. For the following culturing, calcium-free Dulbecco's Modified Eagle Medium (DMEM) including the aforementioned supplements was used. For all cell experiments cells of six donors were pooled. Three donors were female at the age of 39, 52 and 63 years, respectively; three were male at the age of 18, 42 and 44 years, respectively. All experiments were conducted using with cells at passages lower than 6. The osteoblastic phenotype of cells was confirmed by immunocytochemistry staining as detailed below.

4.2.6 Culturing of human fibroblasts (HFIB)

Some experiments were performed with a bought human fibroblasts cell line. The donor of the cell line is a 41 year Caucasian female. For culturing DMEM medium was used, supplemented with 5 % FCS, 1.6 mM L-glutamine, and priomycin.

4.2.7 Cell handling

Cell splitting

Cells were split according to the cell growth rate and density on the culture flasks. For splitting, cells were washed once in PBS consisting of 150 mmol/ l NaCl, 8 mmol/ l Na_2HPO_4, 3 mmol/ l KCl und 1.5 mmol/ l KH_2PO pH 7.4, and removed from the culture dish by incubation for 5 min at 37 °C with 2 ml/ 150 cm^2 trypsin/EDTA solution. By tapping cells were detached from the bottom of the culture flask. An equal volume of medium was added to the cell suspension, centrifuged (10 min, 2500 rpm, 23 °C), the pellet was resuspended in fresh medium and cells were accordingly inoculated. For the experiments we used primary cells that were not splIt more than 4 times and SAOS cells with a passage number under 15. The number of cells was counted using an electronic cell counter (CASY) (sample diluted 1: 200 in CASY®ton) or by counting chamber.

Freezing and unfreezing

Before freezing, cells were treated as described for the splitting process. After being centrifuged the pellet was resuspended with freezing media in a minimum density of 5×10^6 cells/ ml. The aliquots were immediately put in freezing boxes, filled with 1-propanol, and frozen at -80 °C. After 24 hours the cryotubes were transferred in liquid nitrogen.

To unfreeze 5-10 ml of media were pre-warmed in a tube. A cryotube with frozen cells was thawed in a 37 °C water bath. Directly after the cell suspension is unfrosted, it is transferred to the prepared media to dilute the DMSO of the freezing media. After centrifugation for 10 min at 2500 rpm the pellet was resuspended in media and the cells were counted and seeded in culture flasks as required.

4.2.8 Immunohistology protocol

Osteocalcin (OC), fibronectin (FN) and collagen I (CI)

Cells were cultured on a CultureSlide and fixed with 4 % paraformaldehyde (PFA) or ethanol/ methanol (1:1, v/ v). For staining alkaline phosphatase (ALP) NBT/ BCIP Ready-to-use tablets (nitro blue tetrazolium chloride/5-bromo-4-chloro-3-indolylphosphate) were used. Osteocalcin (OC), fibronectin (FN) and collagen I (CI) were detected by immunocytochemistry. First an incubation step of FN and OC with proteinase K (1:1000) was performed over 5 and 20 min, respectively. CI was treated with pepsin (0.4% in 0.01M HCl) for 40 min at 37 °C. After washes in PBS and blocking with Protein Block for 10 min, the primary antibodies directed to FN (1: 4000), OC (1: 50) and CI (1: 25), respectively, were incubated for 30, 90 min, or overnight. For each marker a negative control omitting the primary antibody was included. The cells were washed twice in PBS-Brij (0.01 % Brij in PBS) and then the secondary antibody (rabbit 1: 200) was added for 30 min. After three washes in PBS-Brij, an incubation step with avidin-biotin complex was performed for 30 min. The final washing was done twice in PBS-Brij and once in PBS. The bound avidin-biotin complex was visualized by incubation with the high sensitivity substrate chromogen 3-amino-9-ethylcarbazole for about 15 min. All steps were done at RT, except the incubation of the antibody raised against CI, which was performed at 4 °C. Signal intensity was evaluated using a transmitted-light microscope (AxioVert) equipped with a digital camera (AxioCam ICc3).

Alkaline phosphatase (ALP)

Cells were cultured on CultureSlides and fixed with 4 % paraformaldehyde or ethanol/methanol (volume ratio 1: 1). For staining ALP NBT/BCIP Ready-to-use tablets (nitro blue tetrazolium chloride/5-bromo-4-chloro-3-indolylphosphate) were used. The reaction was stopped by removing the staining solution with PBS. The staining is based on the ALP catalyzed staining reaction. 5-Bromo-4-chloro-3-indolyl phosphate (BCIP) conduces as substrate for ALP, which after dephosphorylation is oxidized by NBT to yield a dark-blue indigo precipitating dye. NBT is thereby reduced to the dark-blue precipitating dye diformazan and serves to intensify the color reaction making the detection more sensitive. Both dyes drop out near by the ALP molecules and stain the area of the ALP positive cells dark purple.[196] For haemalaun staining, the culture slides were incubated 1 min at RT in haemalaun. The reaction was stopped by washing the slides in water. Afterwards the culture slides were directly covered with glycerol-gelatine and fixed with a cover slip. The analysis of the ALP staining result and counting of the ALP positive cells was executed using a transmitted-light microscope and pictures were taken using a digital camera.

Integrin staining

Cells were cultured on CultureSlides that were coated with fibronectin (5 µg/ ml) for 30 min. Immunofluorescent integrin staining was conducted after cell fixation by 4 % PFA. Thereafter, cells were washed in PBS, 2 % (w/ v) BSA. The following primary antibodies were used: CD51/61 (1: 100), CD51 (1: 200) and CD61 (1: 100). Incubation time was 90 min at RT, followed by with the addition of the secondary Alexa-488-conjugated goat-anti-mouse IgG for 45 min at RT. For confocal laser scanning microscopy (CLSM), slides were mounted in PBS and fluorescence signal intensity determined. Staining procedures in the presence of the secondary Alexa-488-labeled antibody alone served as controls and resulted in negligible fluorescence signals.

4.2.9 Lactate dehydrogenase (LDH)

Coating of the disks was done as explained for the $\alpha v \beta 3$-adhesion assay. After three washes in PBS disks were blocked with 5 % BSA in PBS for 1 h at 30 °C. For the quantification of cellular viability several standard assays have been developed, for example the Cytotoxicity Detection Kit (LDH). For this test, LDH a stable cytoplasmic enzyme present in all cells is detected in the supernatant by a single measurement at one time point. Cells were seeded at a density of 25,000 cells per disk in DMEM with 15 % FCS for 24 h at 37 °C in 5 % CO_2. Afterwards the supernatant was collected cell free and centrifuged at 250 x g for 10 min at RT. Afterwards the supernatant is transferred in a new multi well plate and incubated with same volume of reaction mixture for 0.5 h shaded at RT. The LDH-catalyzed conversion of lactate to pyruvate reduces NAD^+ to $NADH/ H^+$. In the second step the catalyst (diaphorase) transfers H/ H^+ from $NADH/ H^+$ to the tetrazolium salt INT which is reduced to formazan. When the amount of dead or plasma membrane-damaged cells is increased, the LDH enzyme activity rises and correlates to the amount of formazan formed during the given time period. Therefore the amount of color formed is proportional to the number of lysed cells. The absorption of the water soluble formazan dye was measured at 492 nm in a micro plate reader. For the calibration curve osteoblasts were seeded at different concentrations and treated respectively. As negative control cell cultured only on a plastic multi well plate were used and in the wells of the positive control cells were treated with Triton X-100 (100 µl/ well) at the end of the incubation time for 10 min. Data are given as mean value (± standard deviations) of at least three identical but independent experiments (probes within one assay performed in triplicate).

4.2.10 Evaluation of osteoblast adhesion by chromogenic hexosaminidase test

Coating of the disks was done as explained for the $\alpha v \beta 3$-adhesion assay. After three washes in PBS disks were blocked with 5 % BSA in PBS for 1 h at 30 °C. The chromogenic cell adhesion assay was performed as described by Landegren et al.[197] Cells were seeded at a density of 25,000 cells per disk in DMEM without FCS. After 3 h of adhesion time at 37 °C in 5 % CO_2 cells on top of the disks were washed three times in PBS in order to remove all nonadherent cells. Adherent cells were quantified by detection of the lysomal enzyme hexosaminidase that hydrolysis the substrate p-nitrophenol-N-acetyl-β-D-glucosaminide into N-acetyl-D-glucosamine and p-nitrophenol. The reaction was stopped by adding a stop solution (0.2 M NaOH, 5 mM Ethylendiamintetra-acetat [EDTA]). The amount of colored p-nitrophenol was measured at 405 nm in an ELISA reader. For the calibration curve osteoblasts were seeded at different concentrations and treated similarly. The adhesion strength was determined as percentage of the applied cell number by setting the signal obtained for 25,000 cells by the standard curve to 100 %. Data are given as mean value (± standard deviations) of at least three identical but independent experiments (probes within one assay performed in triplicate).

4.2.11 Fluorescent staining

Osteoblasts were seeded at a density of 25,000 cells per disk. The adherence time was 1 h or 3 h at 37 °C under 5 % CO_2, followed by three washes in PBS. The fluorescence staining was performed by adding fluorescein diacetate (FDA) solution for 10 min. This fluorogen (excitation spectra: 456 nm, emission sprectra: 520- 530 nm) is often used for vital staining of cells, as it fluoresces only when metabolized within the cell.[198,199] The staining reaction was stopped by washes in PBS. Disks were immediately examined using a reflected-light microscope (Discovery.V8 SteREO). Spread cells on each disk were counted in five randomly chosen spots. Experiments were performed in triplicate. Data are given as number of spread cells in per cent (%) of all analyzed cells after 1 and 3 h, respectively.

4.2.12 Video spreading

In cooperation with Prof. Bausch and Dr. Fernandez channels were produced by photolithography as described by Sidorova.[200] In a next step channels were coated o/ N with 100 µM peptide **1** or **2**. After blocking with 5 % BSA for 2 h at 30 °C osteoblasts were inoculated. The whole system was monitored over 6 h at 37 °C by taking every 30 seconds a picture of previous defined areas in the channels. The level of cell spreading was measured by a routine programmed by Maximilian Baust; on 3 independent channels the area of 3 cells was counted and analyzed.

4.2.13 SEM images

The morphology of osteoblasts adherent to RGD coated Ti6Al4V disks after 3 h of culture was analyzed by means of a SEM. To this end, cells were fixed in 1 % (v/ v) glutaraldehyde in phosphate buffer for 24 h at 4 °C, rinsed three times in H_2O, dehydrated in graded alcohol (70 %, 80 %, 90 %, 100 %), critical-point dried with CO_2, and sputter-coated. The disks were examined using a Leo 440i SEM at an accelerating voltage of 10 kV.

4.2.14 Sheep experiment

In the animal experiment monolithic ceramic probes BIOLOX® delta (aluminum oxide-zirconium oxide-composite ceramic) with different bioactive coatings were used. The probes had an identical size of 4 mm radius and 11 mm length, the surface texture was structured (pore size: 200-500 µm, porosity 45 %) or unstructured. On structured probes different techniques for bioactivation were performed: hydroxyapatite coating, plasma treatment, and coating with 100 µM peptide RGDfK-Ahx-SH (RGD) o/N.

PD Dr. Burgkart performed all animal surgeries, the push-out experiments were accomplished at the biomechanical department under supervision of PD Dr. Burgkart, and Susanne Kerschbaumer arranged the histological preparation and the staining.

The experiment was conducted in a veterinary surgical center according to animal studies ethical principles. The veterinary license of the administration of Oberbayern and all further information are documented in the separate "Abschlussbericht Tierversuche D2B 2008" as well as in the "Anlage zum Abschlussbericht Tierversuche D2B 2008".

24 female Merino sheep, 55 to 87 kg, were used. Cefuroxime was given by intravenous injection as a preventive antibiotic. After an endotracheal intubation anesthesia was maintained with an intravenous perfusion of isoflurane. In addition, novalgin and ketoprofen provided an analgetic coverage. Tibia was exposed by a lateral and medial approach. The recipient and reproducible sites were created with a drill under physiologic saline. The ceramic implants were placed by press fit after the cavities have been flushed by sterile physiologic saline to remove bone debris. Each animal received 4 ceramic implants per knee region, as illustrated in Figure 58, the

exact implantation plan can be found in the "Endbericht Verbundprojekt Direct To Bone". Incision was closed in three-layer by standard suture.

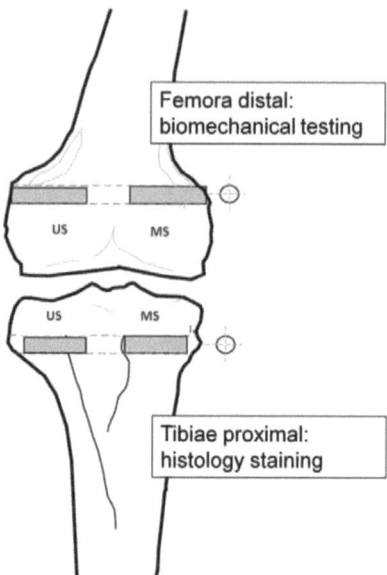

Figure 58: Overview of positioning of ceramic probes in sheep femora and tibiae.

After euthanasia the soft tissue of the femora and tibiae was removed and either used fresh for the push-out tests or transferred to 4 % formalin as preparation of the histological staining.

For the push-out tests the femoral tibiae were investigated in a special testing set-up. Using a self-established adjusting device the probes were exactly orthograde prepared and isogonal positioned in the text fixture. The push-out testing was performed in a standardized way by using a universal testing machine (Wolpert, TZZ 707). The corresponding forces were documented.

The histological staining was performed on the tibial probes. The bone samples containing the ceramic probes were embedded in formalin for 2 days. Because of the high amount of fat in sheep bones the embedding in formalin lasted very long and interruptions of the polymerization in terms of bubbling and air pockets can occur. However, this does not affect the assessment of findings. Further the hardness of the ceramic probes challenged the set-up for sawing the probes, as the time period for one probe took some days.

The probes were separated sagittally and cut in the frontal plane, affixed on glass slides and processed to 140- 80 µm thick microsections. The surfaces were polished by corundum sand paper (1200, 2400, 4000). The staining was performed the self-established staining protocol, all histological evaluations as well as the documentation were done on a Leica DMRB with Zeiss Axiocam camera.

For the interpretation of connective tissue, cellular parts, differentiation grade, bone matrix and quality of bone, as well as bone activity and vitality, the probes were stained selectively. The following evaluation matrix was used:

- Orange/ red: connective tissue, cell nucleus
- Brown: collagen, collagen fibrils
- Brown/ red/ yellow: cartilage, osteoid
- Green: mineralized bone, starting points for mineralization (point-shaped), calcium salt (unstructured)
- light green: old bone
- pale green/black: non vital bone, necrotic bone

4.2.15 Statistical analysis

All data presented in this study are given as mean values (± standard deviations) of at least three experiments. Significant differences between group means were analyzed by the student's t-test. A p-value < 0.05 was considered statistically significant.

LITERATURE

[1] P. F. Gomez, J. A. Morcuende, *Iowa Orthopedic Journal* **2005**, *25*, 25-29.

[2] H. Grundei. Geschichtliche Entwicklung der Endoprothetik und der Fixation durch Spongiosa-Metal, in *Ossäre Integration*, (Eds.: R. Gradinger, H. Gollwitzer), Springer, Heidelberg **2006**, pp. 2-11.

[3] Gördes W., Plötz W. Hüftendoprothetik, in *Orthopädie und Traumatologie*, (Eds.: Hipp E.G., Plötz W., Thiemel G.), Thieme, Stuttgart **2003**, pp. 623-644.

[4] P. A. Ring, *Annals of The Royal College of Surgeons of England* **1971**, *48*, 344-355.

[5] Website of Orthodynamics. 2011. http://www.orthodynamics.de/cms/front_content.php?idcat=23

[6] L. Claes, *Langenbecks Arch Chir* **1987**, *372*, 441-446.

[7] E. Wintermantel, B. Shah-Derler, A. Bruinik, M. Petimermet, J. Blum, S.-W. Ha. Biokompatibilität, in *Medizintechnik - Life Science Engineering*, (Eds.: E. Wintermantel, S.-W. Ha), Springer, Berlin **2009**, pp. 67-112.

[8] P.-I. Branemark, *The osseointegration book: from calvarium to calcaneus*, Quintessenz, Berlin **2006**.

[9] R. K. Schenk, D. Buser, *Periodontol.2000.* **1998**, *17*, 22-35.

[10] E. Wintermantel, S.-W. Ha. Metalle, in *Medizintechnik mit biokompatiblen Werkstoffen und Verfahren*, (Eds.: E. Wintermantel, S.-W. Ha), Springer, Berlin **2002**, pp. 121-147.

[11] X. Liu, P. K. Chu, C. Ding, Mater. Sci. Eng. **2004**, *47*, 49-121.

[12] J. Lincks, B. D. Boyan, C. R. Blanchard, C. H. Lohmann, Y. Liu, D. L. Cochran, D. D. Dean, Z. Schwartz, *Biomaterials* **1998**, *19*, 2219-2232.

[13] S. G. Steinemann, *Periodontol.2000.* **1998**, *17*, 7-21.

[14] M. Long, H. J. Rack, *Biomaterials* **1998**, *19*, 1621-1639.

[15] J. A. Jansen, A. F. von Recum, J. P. van der Waerden, K. de Groot, *Biomaterials* **1992**, *13*, 959-968.

[16] K. Klingbeil. Metallurgische Grundlagen für die gusstechnische Herstellung einer räumlichen Oberflächenstruktur, in *Ossäre Integration*, (Eds.: R. Gradinger, H. Gollwitzer), Springer, Heidelberg **2006**, pp. 46-52.

[17] O. E. Pohler, *Injury* **2000**, *31 Suppl 4*, 7-13.

[18] R. Van Noort, *J Mater Sci* **1987**, *22*, 3801-3811.

[19] J. E. Ellingsen, *Periodontol.2000.* **1998**, *17*, 36-46.

[20] Riley K, FDA approves first ceramic-on-metal total hip replacement system. **2011**. http://www.fda.gov/NewsEvents/Newsroom/PressAnnouncements/ucm259061.htm

[21] H.-J. Schmitz, R. Kettner, S. Eren. Implantatmaterialien, in *Implantologie*, (Eds.: B. Koeck, W. Wagner), Urban&Fischer, München **1999**, pp. 11-30.

[22] J. Chen, S. Mwenifumbo, C. Langhammer, J. P. McGovern, M. Li, A. Beye, W. O. Soboyejo, *J Biomed.Mater Res.B Appl.Biomater.* **2007**, *82*, 360-373.

[23] J. H. Dumbleton, M. T. Manley, A. A. Edidin, *J Arthroplasty* **2002**, *17*, 649-661.

[24] Y. H. Zhu, K. Y. Chiu, W. M. Tang, *J Orthop.Surg.(Hong.Kong.)* **2001**, *9*, 91-99.

[25] W. H. Harris, *Clin.Orthop.Relat Res.* **1995**, 311, 46-53.

[26] H. G. Willert, G. H. Buchhorn. The biology of loosening of hip implants, in *European Instructional Course Lectures*, (Eds.: R. P. Jakob, P. Fulford, F. Horan), The British Editorial Society of Bone and Joint Surgery, London **1999**, pp. 58-82.

[27] Bitzer, E. M., Grobe, T. G., Neusser, S., Schneider, A., Dörning, H., Schwartz, F. W., Barmer GEk Report Krankenhaus. **2010**. http://www.barmer-gek.de/barmer/web/Portale/Presseportal/Subportal/Infothek/Studien-und-Reports/Report-Krankenhaus/Krankenhausreport-2010/PDF-Report-Krankenhaus-2010

[28] Garellick, G., Kärrholm, J., Rogmark, C., Herberts, P., Swedisch Hip Arthroplasty Register. **2009**. http://www.shpr.se/en/Publications/DocumentsReports.aspx

[29] N. Aebli, J. Krebs, G. Davis, M. Walton, M. J. Williams, J. C. Theis, *Spine (Phila Pa 1976.)* **2002**, *27*, 460-466.

[30] N. Aebli, R. Pitto, J. Krebs, *Schweiz Med Froum* **2005**, *5*, 512-518.

[31] B. Elmengaard, J. E. Bechtold, K. Soballe, *Biomaterials* **2005**, *26*, 3521-3526.

[32] U. Lucht, *Acta Orthop.Scand.* **2000**, *71*, 433-439.

[33] T. J. Puolakka, K. J. Pajamaki, P. J. Halonen, P. O. Pulkkinen, P. Paavolainen, J. K. Nevalainen, *Acta Orthop.Scand.* **2001,** *72,* 433-441.

[34] A. Shekaran, A. J. Garcia, *J Biomed.Mater.Res.A* **2011,** *96,* 261-272.

[35] P. A. Hill, *Br.J Orthod.* **1998,** *25,* 101-107.

[36] D. J. Hadjidakis, I. I. Androulakis, *Ann.N.Y.Acad.Sci* **2006,** *1092,* 385-396.

[37] H. Zeng, K. K. Chittur, W. R. Lacefield, *Biomaterials* **1999,** *20,* 377-384.

[38] C. P. Adler. Bones and bone tissue: normal anatomy and histology, in *Bone Diseases*, Springer, New York **2000**, pp. 1-30.

[39] S. C. Manolagas, *Endocr.Rev.* **2000,** *21,* 115-137.

[40] R. L. Jilka, R. S. Weinstein, T. Bellido, A. M. Parfitt, S. C. Manolagas, *J Bone Miner.Res.* **1998,** *13,* 793-802.

[41] P. Ducy, T. Schinke, G. Karsenty, *Science* **2000,** *289,* 1501-1504.

[42] J. Vilamitjana-Amedee, R. Bareille, F. Rouais, A. I. Caplan, M. F. Harmand, *In Vitro Cell Dev.Biol.Anim* **1993,** *29A,* 699-707.

[43] W. J. Grzesik, P. G. Robey, *J.Bone Miner.Res.* **1994,** *9,* 487-496.

[44] P. V. Bodine, B. S. Komm, *Bone* **1999,** *25,* 535-543.

[45] J. E. Aubin, F. Liu. The osteoblast lineage, in *Principle of bone biology*, (Eds.: J. P. Bilezikian, L. G. Raisz, G. A. Rodan), Academic Press, San Diego **1996**, pp. 51-68.

[46] J. E. Puzas. Osteoblast Cell Biology-Lineage and Function, in *Primer on the metabolic bone diseases and disorders of mineral metabolism*, (Ed.: M. J. Favus), Lippincott-Raven, Philadelphia **1996**, pp. 11-16.

[47] M. R. Khan, N. Donos, V. Salih, P. M. Brett, *Bone* **2011,** *50,* 1-8.

[48] S. P. Bruder, D. J. Fink, A. I. Caplan, *J Cell Biochem.* **1994,** *56,* 283-294.

[49] E. M. Younger, M. W. Chapman, *J Orthop.Trauma* **1989,** *3,* 192-195.

[50] R. G. LeBaron, K. A. Athanasiou, *Tissue Eng* **2000,** *6,* 85-103.

[51] M. López-Garíca, H. Kessler. Stimulation of Bone Growth on Implant by Integrin Ligands, in *Handbook of Biomineralization*, (Eds.: M. Epple, E. Bäuerlein), Wiley-VCH, Weinheim **2007**, pp. 109-126.

[52] E. Zamir, B. Geiger, *J Cell Sci* **2001,** *114,* 3583-3590.

[53] S. Drotleff, U. Lungwitz, M. Breunig, A. Dennis, T. Blunk, J. Tessmar, A. Göpferich, *Euro J Pharma Biopharma.* **2004**, *58,* 385-407.

[54] E. Sackmann, R. F. Bruinsma, *Chemphyschem.* **2002**, *3,* 262-269.

[55] B. Alberts, A. Johnson, J. Lewis, M. Raff, K. Roberts, P. Walter. Cell Junctions, Cell Adhesion, and the Extracellular Matrix, in *Molecular Biology of the Cell,* (Eds.: B. Alberts, A. Johnson, J. Lewis, M. Raff, K. Roberts, P. Walter), Garland Science, New York **2002**, pp. 1169-1180.

[56] R. O. Hynes, *Cell* **2002**, *110,* 673-687.

[57] U. Hersel, C. Dahmen, H. Kessler, *Biomaterials* **2003**, *24,* 4385-4415.

[58] M. D. Pierschbacher, E. Ruoslahti, *Nature* **1984**, *309,* 30-33.

[59] T. Xiao, J. Takagi, B. S. Coller, J. H. Wang, T. A. Springer, *Nature* **2004**, *432,* 59-67.

[60] A. van der Flier, A. Sonnenberg, *Cell Tissue Res.* **2001**, *305,* 285-298.

[61] J. D. Humphries, A. Byron, M. J. Humphries, *J Cell Sci.* **2006**, *119,* 3901-3903.

[62] J. Emsley, C. G. Knight, R. W. Farndale, M. J. Barnes, R. C. Liddington, *Cell* **2000**, *101,* 47-56.

[63] M. Shimaoka, T. Xiao, J. H. Liu, Y. Yang, Y. Dong, C. D. Jun, A. McCormack, R. Zhang, A. Joachimiak, J. Takagi, J. H. Wang, T. A. Springer, *Cell* **2003**, *112,* 99-111.

[64] T. Goto, M. Yoshinari, S. Kobayashi, T. Tanaka, *Biomed.Mater Eng* **2004**, *14,* 537-544.

[65] S. Miyamoto, H. Teramoto, O. A. Coso, J. S. Gutkind, P. D. Burbelo, S. K. Akiyama, K. M. Yamada, *J Cell Biol.* **1995**, *131,* 791-805.

[66] C. Brakebusch, R. Fässler, *EMBO* **2003**, *22,* 2324-2333.

[67] S. Wiesner, K.-R. Legate, R. Fässler, *Cell Mol.Life Sci* **2005**, *62,* 1081-1099.

[68] M. Vicente-Manzanares, C. K. Choi, A. R. Horwitz, *J Cell Sci.* **2009**, *122,* 199-206.

[69] W. H. Goldmann, R. M. Ezzell, E. D. Adamson, V. Niggli, G. Isenberg, *J Muscle Res.Cell Motil.* **1996**, *17,* 1-5.

[70] G. Gronowicz, M. B. McCarthy, *J Orthop.Res.* **1996**, *14,* 878-887.

[71] B. H. Luo, T. A. Springer, *Curr.Opin.Cell Biol.* **2006**, *18,* 579-586.

[72] J. Zhu, B. Boylan, B. H. Luo, P. J. Newman, T. A. Springer, *J Biol.Chem.* **2007**, *282*, 11914-11920.

[73] M. C. Siebers, P. J. ter Brugge, X. F. Walboomers, J. A. Jansen, *Biomaterials* **2005**, *26*, 137-146.

[74] S. D. Redick, D. L. Settles, G. Briscoe, H. P. Erickson, *J Cell Biol.* **2000**, *149*, 521-527.

[75] K. O. Simon, E. M. Nutt, D. G. Abraham, G. A. Rodan, L. T. Duong, *J Biol.Chem.* **1997**, *272*, 29380-29389.

[76] K. Anselme, P. Linez, M. Bigerelle, D. Le Maguer, A. Le Maguer, P. Hardouin, H. F. Hildebrand, A. Iost, J. M. Leroy, *Biomaterials* **2000**, *21*, 1567-1577.

[77] V. W. Engleman, G. A. Nickols, F. P. Ross, M. A. Horton, D. W. Griggs, S. L. Settle, P. G. Ruminski, S. L. Teitelbaum, *J Clin.Invest* **1997**, *99*, 2284-2292.

[78] R. K. Sinha, R. S. Tuan, *Bone* **1996**, *18*, 451-457.

[79] D. Goltzman, *Nat.Rev.Drug Discovery* **2002**, *1*, 784-796.

[80] D. E. Hughes, D. M. Salter, S. Dedhar, R. Simpson, *J.Bone Miner.Res.* **1993**, *8*, 527-533.

[81] S. Gronthos, K. Stewart, S. E. Graves, S. Hay, P. J. Simmons, *J Bone Miner.Res.* **1997**, *12*, 1189-1197.

[82] C. T. Brighton, S. M. Albelda, *J Orthop.Res.* **1992**, *10*, 766-773.

[83] S. Rammelt, T. Illert, S. Bierbaum, D. Scharnweber, H. Zwipp, W. Schneiders, *Biomaterials* **2006**, *27*, 5561-5571.

[84] K. M. Hennessy, W. C. Clem, M. C. Phipps, A. A. Sawyer, F. M. Shaikh, S. L. Bellis, *Biomaterials* **2008**, *29*, 3075-3083.

[85] J. E. Ho, T. A. Barber, A. S. Virdi, D. R. Sumner, K. E. Healy, *J Biomed.Mater.Res.A* **2007**, *81*, 720-727.

[86] D. M. Ferris, G. D. Moodie, P. M. Dimond, C. W. Gioranni, M. G. Ehrlich, R. F. Valentini, *Biomaterials* **1999**, *20*, 2323-2331.

[87] M. Friedlander, C. L. Theesfeld, M. Sugita, M. Fruttiger, M. A. Thomas, S. Chang, D. A. Cheresh, *Proc.Natl.Acad.Sci U.S.A* **1996**, *93*, 9764-9769.

[88] M. S. Goligorsky, H. Kessler, V. I. Romanov, *Nephrol.Dial.Transplant.* **1998**, *13*, 254-263.

[89] Z. Yun, D. G. Menter, G. L. Nicolson, *Cancer Res.* **1996**, *56*, 3103-3111.

[90] W. Pompe, H. Worch, M. Epple, W. Firess, M. Gelinsky, P. Greil, U. Hempel, D. Scharnweber, K. Schulte, Mater. Sci. Eng., A **2003**, *362*, 40-60.

[91] A. Bagno, A. Piovan, M. Dettin, A. Chiarion, P. Brun, R. Gambaretto, G. Fontana, C. Di Bello, G. Palu, I. Castagliuolo, *Bone* **2007**, *40*, 693-699.

[92] A. Bagno, A. Piovan, M. Dettin, P. Brun, R. Gambaretto, G. Palu, C. Di Bello, I. Castagliuolo, *Bone* **2007**, *41*, 704-712.

[93] M. Lampin, C. Warocquier, C. Legris, M. Degrange, M. F. Sigot-Luizard, *J Biomed.Mater.Res.* **1997**, *36*, 99-108.

[94] C. Dahmen, J. Auernheimer, A. Meyer, A. Enderle, S. L. Goodman, H. Kessler, *Angew.Chem.Int.Ed Engl.* **2004**, *43*, 6649-6652.

[95] B. Nebe, B. Finke, F. Luthen, C. Bergemann, K. Schroder, J. Rychly, K. Liefeith, A. Ohl, *Biomol.Eng* **2007**, *24*, 447-454.

[96] K. Anselme, M. Bigerelle, B. Noel, E. Dufresne, D. Judas, A. Iost, P. Hardouin, *J Biomed.Mater.Res.* **2000**, *49*, 155-166.

[97] S. Giljean, A. Ponche, M. Bigerelle, K. Anselme, *J Biomed.Mater.Res.A* **2010**, *94*, 1111-1123.

[98] K. Anselme, M. Bigerelle, *Biomaterials* **2006**, *27*, 1187-1199.

[99] R. Lange, F. Lüthen, U. Beck, J. Rychly, A. Baumann, B. Nebe, *Biomol.Eng* **2002**, *19*, 255-261.

[100] F. Lüthen, R. Lange, P. Becker, J. Rychly, U. Beck, J. G. Nebe, *Biomaterials* **2005**, *26*, 2423-2440.

[101] E. Lieb, M. Hacker, J. Tessmar, L. A. Kunz-Schughart, J. Fiedler, C. Dahmen, U. Hersel, H. Kessler, M. B. Schulz, A. Gopferich, *Biomaterials* **2005**, *26*, 2333-2341.

[102] B. Chehroudi, T. R. Gould, D. M. Brunette, *J Biomed.Mater.Res.* **1989**, *23*, 1067-1085.

[103] A. Rezania, K. E. Healy, *J Orthop.Res.* **1999**, *17*, 615-623.

[104] G. Sagvolden, I. Giaever, E. O. Pettersen, J. Feder, *Proc.Natl.Acad.Sci U.S.A* **1999**, *96*, 471-476.

[105] A. Rezania, C. H. Thomas, K. E. Healy, *Ann.Biomed.Eng* **1997**, *25*, 190-203.

[106] T. Blunk, A. Göpferich, J. Tessmar, *Biomaterials* **2003**, *24*, 4335.

[107] S. Drotleff, U. Lungwitz, M. Breunig, A. Dennis, T. Blunk, J. Tessmar, A. Göpferich, *Eur.J Pharm.Biopharm.* **2004**, *58*, 385-407.

[108] M. Aumailley, M. Gurrath, G. Müller, J. Calvete, R. Timpl, H. Kessler, *FEBS Lett.* **1991**, *291*, 50-54.

[109] C. Mas-Moruno, F. Rechenmacher, H. Kessler, *Anticancer Agents Med.Chem.* **2010**, *10*, 753-768.

[110] M. Kantlehner, P. Schaffner, D. Finsinger, J. Meyer, A. Jonczyk, B. Diefenbach, B. Nies, G. Hölzemann, S. L. Goodman, H. Kessler, *Chembiochem.* **2000**, *1*, 107-114.

[111] C. D. Reyes, T. A. Petrie, K. L. Burns, Z. Schwartz, A. J. García, *Biomaterials* **2007**, *28*, 3228-3235.

[112] T. Miyata, M. S. Conte, L. A. Trudell, D. Mason, A. D. Whittemore, L. K. Birinyi, *J Surg.Res.* **1991**, *50*, 485-493.

[113] J. A. Neff, K. D. Caldwell, P. A. Tresco, *J Biomed.Mater.Res.* **1998**, *40*, 511-519.

[114] P. Schaffner, M. M. Dard, *Cell Mol.Life Sci* **2003**, *60*, 119-132.

[115] M. Gurrath, G. Muller, H. Kessler, M. Aumailley, R. Timpl, *Eur.J Biochem.* **1992**, *210*, 911-921.

[116] R. Haubner, D. Finsinger, H. Kessler, *Angew.Chem.Int.Ed* **1997**, *109*, 1440-1456.

[117] M. Pfaff, K. Tangemann, B. Muller, M. Gurrath, G. Muller, H. Kessler, R. Timpl, J. Engel, *J Biol.Chem.* **1994**, *269*, 20233-20238.

[118] A. Howe, A. E. Aplin, S. K. Alahari, R. L. Juliano, *Curr.Opin.Cell Biol.* **1998**, *10*, 220-231.

[119] M. A. Dechantsreiter, E. Planker, B. Mathä, E. Lohof, G. Hölzemann, A. Jonczyk, S. L. Goodman, H. Kessler, *J Med.Chem.* **1999**, *42*, 3033-3040.

[120] R. Haubner, R. Gartias, B. Diefenbach, S. L. Goodman, A. Jonczyk, H. Kessler, *J.Am.Chem.Soc.* **1996**, *118*, 7461-7472.

[121] J. Auernheimer, C. Dahmen, U. Hersel, A. Bausch, H. Kessler, *J.Am.Chem.Soc.* **2005**, *127*, 16107-16110.

[122] M. Kantlehner, D. Finsinger, J. Meyer, P. Schaffner, A. Jonczyk, B. Diefenbach, B. Nies, H. Kessler, *Angew.Chem.Int.Ed* **1999**, *38*, 560-562.

[123] J. Auernheimer, H. Kessler, *Bioorg.Med.Chem.Lett.* **2006**, *16*, 271-273.

[124] D. M. Brunette, P. Tengvall, M. Textor, P. Thomsen, *Titanium in medicine*, Springer, Berlin **2001**.

[125] B. Elmengaard, J. E. Bechtold, K. Soballe, *J Biomed.Mater.Res.A* **2005**, *75*, 249-255.

[126] H. C. Kroese-Deutman, D. J. van den, P. H. Spauwen, J. A. Jansen, *Tissue Eng* **2005**, *11*, 1867-1875.

[127] C. Matschegewski, S. Staehlke, R. Loeffler, R. Lange, F. Chai, D. P. Kern, U. Beck, B. J. Nebe, *Biomaterials* **2010**, *31*, 5729-5740.

[128] J. M. Gold, M. Schmidt, S. G. Steinemann, *Helv.Phys.Acta* **1989**, *62*, 246-249.

[129] M. Gnauck, E. Jaehne, T. Blaettler, S. Tosatti, M. Textor, H. J. Adler, *Langmuir* **2007**, *23*, 377-381.

[130] J. Auernheimer, D. Zukowski, C. Dahmen, M. Kantlehner, A. Enderle, S. L. Goodman, H. Kessler, *Chembiochem.* **2005**, *6*, 2034-2040.

[131] V. C. Hirschfeld-Warneken, M. Arnold, A. Cavalcanti-Adam, M. López-Garíca, H. Kessler, J. P. Spatz, *Eur.J Cell Biol.* **2008**, *87*, 743-750.

[132] G. L. Kenausis, J. Vörös, D. L. Elbert, N. Huang, R. Hofer, L. Ruiz-Taylor, M. Textor, J. A. Hubbell, N. D. Spencer, *J.Phys.Chem.* **2000**, *104*, 3298-3309.

[133] U. Magdolen, J. Auernheimer, C. Dahmen, J. Schauwecker, H. Gollwitzer, J. Tubel, R. Gradinger, H. Kessler, M. Schmitt, P. Diehl, *Int.J.Mol.Med.* **2006**, *17*, 1017-1021.

[134] M. Schuler, G. R. Owen, D. W. Hamilton, M. de Wild, M. Textor, D. M. Brunette, S. Tosatti, *Biomaterials* **2006**, *27*, 4003-4015.

[135] M. Arnold, E. A. Cavalcanti-Adam, R. Glass, J. Blümmel, W. Eck, M. Kantlehner, H. Kessler, J. P. Spatz, *Chemphyschem.* **2004**, *5*, 383-388.

[136] A. Rezania, K. E. Healy, *J Biomed.Mater.Res.* **2000**, *52*, 595-600.

[137] A. Rezania, R. Johnson, A. R. Lefkow, K. E. Healy, *Langmuir* **1999**, *15*, 6931-6939.

[138] M. D. Ward, M. Dembo, D. A. Hammer, *Ann.Biomed.Eng* **1995**, *23*, 322-331.

[139] D. J. Mooney, R. Langer, D. E. Ingber, *J Cell Sci* **1995**, *108*, 2311-2320.

[140] T. Matsuura, R. Hosokawa, K. Okamoto, T. Kimoto, Y. Akagawa, *Biomaterials* **2000**, *21*, 1121-1127.

[141] K. Okamoto, T. Matsuura, R. Hosokawa, Y. Akagawa, *J Dent.Res.* **1998**, *77*, 481-487.

[142] B. A. Dalton, C. D. McFarland, P. A. Underwood, J. G. Steele, *J Cell Sci* **1995**, *108*, 2083-2092.

[143] A. A. Sawyer, K. M. Hennessy, S. L. Bellis, *Biomaterials* **2007**, *28*, 383-392.

[144] A. Ben Ze'ev, *Biochim.Biophys.Acta* **1985**, *780*, 197-212.

[145] P. M. Davidson, O. Fromigue, P. J. Marie, V. Hasirci, G. Reiter, K. Anselme, *J Mater.Sci.-Mater.Med.* **2010**, *21*, 939-946.

[146] K. Anselme, A. Ponche, M. Bigerelle, *Proc.Inst.Mech.Eng H.* **2010**, *224*, 1487-1507.

[147] D. Ferrera, S. Poggi, C. Biassoni, G. R. Dickson, S. Astigiano, O. Barbieri, A. Favre, A. T. Franzi, A. Strangio, A. Federici, P. Manduca, *Bone* **2002**, *30*, 718-725.

[148] P. G. Robey, J. D. Termine, *Calcif.Tissue Int.* **1985**, *37*, 453-460.

[149] P. D. Delmas, R. Eastell, P. Garnero, M. J. Seibel, J. Stepan, *Osteoporos.Int.* **2000**, *11 Suppl 6*, S2-17.

[150] B. Saldamli, In vitro Untersuchungen zur Biokompatibilität von Titan nach Sauerstoff-Plasma-Immersions-Ionenimplantation, Dissertation. Universität Basel, **2008**.

[151] U. Meyer, A. Buchter, H. P. Wiesmann, U. Joos, D. B. Jones, *Eur.Cell Mater* **2005**, *9*, 39-49.

[152] T. A. Owen, M. Aronow, V. Shalhoub, L. M. Barone, L. Wilming, M. S. Tassinari, M. B. Kennedy, S. Pockwinse, J. B. Lian, G. S. Stein, *J Cell Physiol* **1990**, *143*, 420-430.

[153] H. Marques da Silva, M. Mateescu, C. Damia, E. Champion, G. Soares, K. Anselme, *Colloids Surf., B* **2010**, *80*, 138-144.

[154] P. Bianco, M. Riminucci, S. Gronthos, P. G. Robey, *Stem Cells* **2001**, *19*, 180-192.

[155] E. Eklou-Kalonji, I. Denis, M. Lieberherr, A. Pointillart, *Cell Tissue Res.* **1998**, *292*, 163-171.

[156] G. A. Rodan, M. Noda, *Crit Rev.Eukaryot.Gene Expr.* **1991**, *1*, 85-98.

[157] S. B. Doty, B. H. Schofield, *Prog.Histochem.Cytochem.* **1976**, *8*, 1-38.

[158] K. Anselme, *Osteoporos.Int.* **2011**, *22*, 2037-2042.

[159] A. Ponche, M. Bigerelle, K. Anselme, *Proc.Inst.Mech.Eng H.* **2010**, *224*, 1471-1486.

[160] A. K. Shah, R. K. Sinha, N. J. Hickok, R. S. Tuan, *Bone* **1999**, *24*, 499-506.

[161] G. E. Pecora, R. Ceccarelli, M. Bonelli, H. Alexander, J. L. Ricci, *Implant.Dent.* **2009**, *18*, 57-66.

[162] E. S. Gadelmawla, M. M. Koura, T. M. A. Maksoud, I. M. Elewa, H. H. Soliman, *J.Mat.Pro.Tech.* **2002**, *123*, 133-145.

[163] T. Sun, L. Feng, X. Gao, L. Jiang, *Acc.Chem.Res.* **2005**, *38*, 644-652.

[164] D. E. Ingber, *J Cell Sci* **2003**, *116*, 1397-1408.

[165] I. Degasne, M. F. Basle, V. Demais, G. Hure, M. Lesourd, B. Grolleau, L. Mercier, D. Chappard, *Calcif.Tissue Int.* **1999**, *64*, 499-507.

[166] J. C. Keller, C. M. Stanford, J. P. Wightman, R. A. Draughn, R. Zaharias, *J Biomed.Mater.Res.* **1994**, *28*, 939-946.

[167] B. Setzer, M. Bachle, M. C. Metzger, R. J. Kohal, *Biomaterials* **2009**, *30*, 979-990.

[168] P. M. Brett, J. Harle, V. Salih, R. Mihoc, I. Olsen, F. H. Jones, M. Tonetti, *Bone* **2004**, *35*, 124-133.

[169] D. W. Hamilton, D. M. Brunette, *Biomaterials* **2007**, *28*, 1806-1819.

[170] K. Anselme, M. Bigerelle, *Acta Biomater.* **2005**, *1*, 211-222.

[171] K. Anselme, M. Bigerelle, *J.Mater.Sci.-Mater.Med.* **2006**, *17*, 471-479.

[172] Z. Schwartz, C. H. Lohmann, J. Oefinger, L. F. Bonewald, D. D. Dean, B. D. Boyan, *Adv.Dent.Res.* **1999**, *13*, 38-48.

[173] J. Auernheimer, R. Haubner, M. Schottelius, H. J. Wester, H. Kessler, *Helv Chim Acta* **2006**, *89*, 833-840.

[174] S. L. Goodman, G. Hölzemann, G. A. Sulyok, H. Kessler, *J Med.Chem.* **2002**, *45*, 1045-1051.

[175] H. Spreitzer, *Österreichische Apothekerzeitung* **2008**, *22*, 1136-1137.

[176] S. J. Jones, A. Boyde, *Cell Tissue Res.* **1977**, *184*, 179-193.

[177] C. Mas-Moruno, P. Dorfner, F. Manzenrieder, S. Neubauer, U. Reuning, R. Burgkart, H. Kessler, *J.Biomed.Mater.Res.* **2012**, *accepted*.

[178] R. G. Duggleby, *Biochem Med Meta Bio* **1988**, *40*, 204-212.

[179] T. Decker, M. L. Lohmann-Matthes, *J.Immunol.Methods* **1988**, *115*, 61-69.

[180] P. C. Brooks, A. M. Montgomery, M. Rosenfeld, R. A. Reisfeld, T. Hu, G. Klier, D. A. Cheresh, *Cell* **1994**, *79*, 1157-1164.

[181] A. Abdollahi, D. W. Griggs, H. Zieher, A. Roth, K. E. Lipson, R. Saffrich, H. J. Grone, D. E. Hallahan, R. A. Reisfeld, J. Debus, A. G. Niethammer, P. E. Huber, *Clin.Cancer Res.* **2005**, *11*, 6270-6279.

[182] D. D. Deligianni, N. Katsala, S. Ladas, D. Sotiropoulou, J. Amedee, Y. F. Missirlis, *Biomaterials* **2001**, *22*, 1241-1251.

[183] P. J. ter Brugge, S. Dieudonne, J. A. Jansen, *J.Biomed.Mater.Res.* **2002**, *61*, 399-407.

[184] M. Pegueroles, C. Aparicio, M. Bosio, E. Engel, F. J. Gil, J. A. Planell, G. Altankow, *Acta Biomater.* **2010**, *6*, 291-301.

[185] R. F. Vogt, Jr., D. L. Phillips, L. O. Henderson, W. Whitfield, F. W. Spierto, *J Immunol.Methods* **1987**, *101*, 43-50.

[186] L. Feng, C. Z. Hu, J. D. Andrade, *J Microsc.* **1988**, *152*, 811-816.

[187] T. Kreis, R. Vale, *Guidebook to the extracellular matrix and adhesion proteins*, (Eds.: T. Kreis, R. Vale) Oxford University Press, Oxford **1993**.

[188] P. G. Robey, A. L. Boskey, The biochemistry of bone, in *Osteoporosis*, (Eds.: R. Marcus, D. Feldman, J. Kelsey) Academic Press, New York **1996**, pp. 95-183.

[189] N. Duewelhenke, P. Eysel, *Orthopäde* **2007**, *36*, 220-226.

[190] B. Kasemo, J. Gold, *Adv.Dent.Res.* **1999**, *13*, 8-20.

[191] P. Roach, D. Farrar, C. C. Perry, *J.Am.Chem.Soc.* **2005**, *127*, 8168-8173.

[192] S. L. Bellis, *Biomaterials* **2011**, *32*, 4202-4210.

[193] Y. N. Danilov, R. L. Juliano, *Exp.Cell Res.* **1989**, *182*, 186-196.

[194] B. Jeschke, J. Meyer, A. Jonczyk, H. Kessler, P. Adamietz, N. M. Meenen, M. Kantlehner, C. Goepfert, B. Nies, *Biomaterials* **2002**, *23*, 3455-3463.

[195] E. Steinhauser, A. Liebendörfer, A. Enderle, R. Bader, S. Kerschbaumer, T. Brill, R. Busch, R. Gradinger, *Materialprüfung* **2005**, *47*, 197-202.

[196] P. L. Wolf, J. P. Horwitz, J. Vazquez, M. E. Von der, *Enzymologia.* **1968**, *35*, 154-156.

[197] U. Landegren, *J.Immunol.Methods* **1984,** *67,* 379-388.

[198] G. G. Guilbault, D. N. Kramer, *Anal.Chem.* **1965,** *37,* 1219-1221.

[199] E. Peak, I. W. Chalmers, K. F. Hoffmann, *PLoS.Negl.Trop.Dis.* **2010,** *4,* e759.

[200] J. M. Sidorova, N. Li, D. C. Schwartz, A. Folch, R. J. Monnat, *Nature Protocols* **2009,** *4,* 849-861.

yes
i want morebooks!

Buy your books fast and straightforward online - at one of world's fastest growing online book stores! Environmentally sound due to Print-on-Demand technologies.

Buy your books online at
www.get-morebooks.com

Kaufen Sie Ihre Bücher schnell und unkompliziert online – auf einer der am schnellsten wachsenden Buchhandelsplattformen weltweit! Dank Print-On-Demand umwelt- und ressourcenschonend produziert.

Bücher schneller online kaufen
www.morebooks.de

VDM Verlagsservicegesellschaft mbH
Heinrich-Böcking-Str. 6-8   Telefon: +49 681 3720 174   info@vdm-vsg.de
D - 66121 Saarbrücken   Telefax: +49 681 3720 1749   www.vdm-vsg.de

Printed by Books on Demand GmbH, Norderstedt / Germany